Challenging Nature

University of Chicago Geography Research Paper
NUMBER 246

Series Editors:

Michael P. Conzen, Neil Harris, Marvin W. Mikesell, Gerald D. Suttles

Titles published in the Geography Research Papers series prior to 1992 and still in print are now distributed by the University of Chicago Press. For a list of available titles, please see the end of the book. The University of Chicago Press commenced publication of the Geography Research Papers series in 1992 with number 233.

Challenging Nature

Local Knowledge, Agroscience, and
Food Security in Tanga Region, Tanzania

PHILIP W. PORTER

THE UNIVERSITY OF CHICAGO PRESS   CHICAGO AND LONDON

PHILIP W. PORTER is emeritus professor of geography at the University of Minnesota.

The University of Chicago Press, Chicago 60637
The University of Chicago Press, Ltd., London
© 2006 by The University of Chicago
All rights reserved. Published 2006
Printed in the United States of America

15 14 13 12 11 10 09 08 07 06   5 4 3 2 1

ISBN: 0-226-67580-7 (cloth)

Library of Congress Cataloging-in-Publication Data

Porter, Philip Wayland, 1928–
     Challenging nature : local knowledge, agroscience, and food security in Tanga region,
     Tanzania / Philip W. Porter.
          p.   cm. — (University of Chicago geography research paper; no. 246)
     Includes bibliographical references and index.
     ISBN 0-226-67580-7 (cloth : alk. paper)
          1. Agriculture—Environmental aspects—Tanzania—Tanga Region.   2. Food supply—
     Tanzania—Tanga Region.   3. Rural development—Tanzania—Tanga Region.
     4. Sustainable   development—Tanzania—Tanga Region.   I. Title.   II. Series.
     S589.76.T34P67   2006
     338.1′0967′822—dc22

                                                                          2005014062

Contents

Acknowledgments

This study is the result of four long-term research programs in East Africa whose purpose has been to understand the environment people use and how people themselves understand their environment and use it to fashion livelihoods. The first research program was an effort to model energy-water-crop relationships and was begun in the early 1960s in connection with the Culture and Ecology in East Africa project, directed by Dr. Walter R. Goldschmidt. It was supported by grants from the National Science Foundation (no. G-11713) and the US Public Health Service (no. MH-74079). A second program involved field research as part of the Agrometeorological Survey of Tanga Region and was undertaken in 1972, when I was a research professor at the University of Dar es Salaam, supported by the Rockefeller Foundation. In 1978–1979, I spent a sabbatical year in Kenya at the Kenya Agricultural Research Institute and associated field research stations. This research was supported in part by the National Science Foundation (no. NSF/SOC-78-15723). Finally, in 1992–1993, I returned to Tanzania, where I was associated with the Institute of Resource Assessment, University of Dar es Salaam, and with the Mlingano Agricultural Research Institute in Tanga Region. I was able to revisit the areas where we worked in 1972. This research was supported in part by a sabbatical furlough from the University of Minnesota, the Fulbright Africa Regional Research Program (no. 92-67419), and the National Science Foundation (no. SES-9210189). I am grateful to these organizations for their support.

My gratitude and indebtedness to individuals could make a long list. I restrict it mainly to two groups: those who have helped me understand energy-water-crop relations and those who have helped me understand the people of Tanga Region and their ways of living.

Lunch break on the mountain road near Vugiri, Korogwe District. Patricia Porter and Pitio Ndyeshumba were largely responsible for the successful completion of the research in 1992–1993. Photo taken 27 February 1993.

The model of energy-water-crop relations has been aided in major ways by Les Maki, Kent Orgain, John Corbett, Greg Flay, Ian Stewart, and Fred Wang'ati. Many people aided my work in Tanzania and Tanga Region. In Dar es Salaam, in 1972, Adolfo Mascarenhas, Dietrich O. Luoga, Stephen Kajula, and Joel Musyani helped me greatly. Student researchers from the University of Dar es Salaam (including the campus at Morogoro) made a significant contribution. I thank R. I. Chambuya, J. A. A. Genda, R. B. Hoza, A. S. Maringo, A. R. Masawe, Ghadi H. Mbwilo, and Alli Shabani. In 1992–1993, David Mees, Christine Djondo, and Roy Southworth provided much help. At the Meteorological Department, I am grateful to Dr. Mhita, Messrs. A. M. Mbuma, Simon Sulle, and Shishira; at the National Archives, Mr. Mkapa; at the University of Dar es Salaam, Dr. Idris Kilula, Dr. James Ngana, Abel Kapele, Peter Tilia, and Charles and Dinah Gasarasi. In Tanga Region, many people were helpful. At the Mlingano Agricultural Research Institute, I was helped by Dr. Samuel E. Mugogo, Maulidi Chanyoya, Philip Kips, Esther Masanzata, Hamisi Shabani, Juven P. Magoggo, and Ms. Karumuna. At the Ubwari Field Station (Muheza), Tony and Dorothy Wilkes, Martin Alilio, James Wakibara, and

Dr. Leonard Mbwera were supportive in numerous ways. I also thank
David Schienman, Tanga, and Moshi O. Mwinyimvua, head of the Tanga
Region Archives. The greatest debts are to Pitio M. B. Ndyeshumba, who
organized and conducted interviews in all the revisited villages, and to my
wife, Pat, who has helped at every step of the way. Finally, I thank the people of Tanga Region who patiently answered our questions and explained
their engagement with the world they use.

For helpful criticism and comment on the completed manuscript, I
am very grateful to James Giblin (historian, University of Iowa), Marcia
Wright (historian, Columbia University), and four former colonial administrators who were in Handeni District in the early 1950s. These are
Randal Sadleir, David Brokensha (anthropologist), Bernard Riley (geographer), and John Ainley (agriculturalist). I am also grateful to Dianne
Rocheleau (Clark University), Marvin Mikesell (University of Chicago),
and three highly conscientious anonymous reviewers. The anonymous reviewers provided many perceptive comments and criticisms that helped
me, particularly regarding the structure of the presentation and in placing
the study within different research contexts. For their thorough and useful editorial suggestions I thank Janet E. Holmén (Karolinska Institute)
and Pamela J. Bruton. I also thank editor Christie Henry (University of
Chicago Press) for her consideration and warm support in bringing this
book to publication.

CHAPTER ONE

Research Questions and Conceptual Contexts

Introduction to the Area and the Research Questions

Agriculture is one of history's greatest inventions and evidence of humankind's intelligence and ingenuity in matching the management of crops in fields with the environment being used. Agriculture is a complex system. Richard Norgaard argued that all complex systems emerge through interactions among many elements over long periods (1994: 81). In many cases they result from a coevolutionary process between a *social* system and its *environmental* system. Consider this statement: "Agriculture began between five and ten thousand years ago when there were approximately five million people in the world. Population doubled eight times, increasing to about 1.6 billion people by the middle of the nineteenth century" (Norgaard 1994: 40). All this was accomplished by people interacting with their local environment in adaptive and, for the most part, sustainable ways. The result was a mosaic of thousands of livelihood systems throughout the world. Each tile in the mosaic was managed by people with local knowledge of themselves and their place. The industrial revolution, European colonial expansion, and agriculture based on petrochemicals changed all that, however, and led to the present world system.

One tile in the mosaic of African agricultural systems can be examined historically in an interesting area that stretches from the Indian Ocean coast westward through Tanga Region, Tanzania, a distance of about

250 km, in an area about the size of Maryland. At the coast, there is a clear bimodal rainfall regime; that is, there are two distinct rainy seasons each year. On the western margins of Tanga Region, there is only one rainy season. Between these two places, there is a gradual transition in the amount, variability, and seasonal distribution of rainfall. In figuring out how to be successful in their agriculture, farmers must take these various aspects of rainfall into account.

There is a good tradition of using transects to study variation across space. Indeed, an interesting transect study was done in Tanga Region across the western Usambara Mountains by Joop Heijnen and Robert W. Kates (1974). Their research examined moisture availability, drought hazards, and local adjustments, topics that also concern us in this study. They found that more favored environments give their residents a greater adaptive capacity to cope with drought hazards.

Lowland Tanga was chosen for this study because of its history of famine and food shortages and because of its good network of rainfall stations with long records (Mörth 1971). The stations form a transect from Moa in the northeast to Mgera in the southwest, from a bimodal peak in rainfall distribution in Moa to a unimodal peak in Mgera (fig. 1.1). Note that in Moa, on the Indian Ocean coast, January and February are drought months[1] and that there are two peaks in precipitation, although admittedly the November peak is not pronounced. If one traveled northward into southeastern Kenya, one would find the November rains to be much greater. At Mgera, on the southwestern margins of Tanga Region, there is only one peak in rainfall (April). June, July, August, and September are drought months.

Tanga Region has a long history of droughts, floods, food shortages, famines, and social and economic disruption. It is an area of persistent rural poverty. Individuals, households, and villages are vulnerable to many risks, which can result in food shortages, threats to health and to the ability to perform labor, and impoverishment. What do the farmers understand about the environment they use? They have been farming here for hundreds of years. Why have they had such a difficult time and experienced food shortages and social disruption so frequently? Is it a matter of a poorly endowed and unreliable environment, or are there other features of their history and economy that account for the disappointing performance of Tanga's farmers?

By historical happenstance, the region has an excellent and long rainfall record because of the presence of sisal estates, begun in German colonial times in the early 20th century. It has been possible to compile a nearly

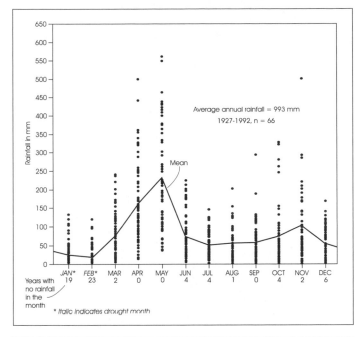

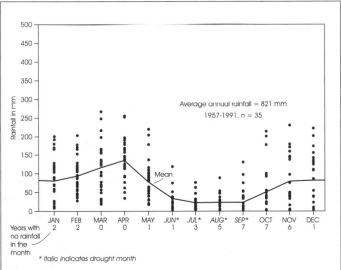

FIG. I.I. Distribution of monthly rainfall at (*top*) Moa (Mtotohovu rainfall station, 9439006) and (*bottom*) Mgera (rainfall station 9537001). (Source: Data from Tanzanian Meteorological Department)

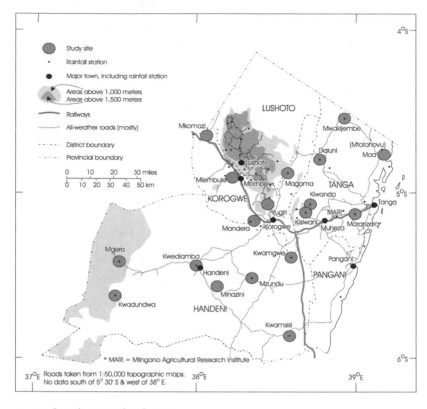

FIG. 1.2. Location map of study areas.

complete record of daily rainfall values from 1926 to 1992 for 25 rainfall stations spread across Tanga Region.[2] The detailed daily rainfall record provides a basis for examining what farmers have done day by day in their fields, as they tried to cope with variable and unknowable weather.

I became interested in trying to see how farmers understood and acted with respect to their ambiguous rainfall resource along this climatic gradient in 1972. In that year, I worked in Tanga Region with a group of students from the University of Dar es Salaam and its agricultural campus (now Konoine University). Our group interviewed 250 farmers in 18 villages spread across Tanga Region, mainly in lowland areas (fig. 1.2). We interviewed a randomly chosen sample of farmers in each village (usually about 50 percent of household heads). We asked questions about life history, environmental and agricultural change, crops grown, timing of planting, environmental assessment, agricultural practices, labor, land, livestock, food supply, water supply, marketing, experience and training, as well as attitudes toward nature and coping measures farmers used when there was drought. We mapped 201 fields and took 50 sets of soil samples. We interviewed knowledgeable local people about soils—the names, characteristics, typical topographic locations, and uses.

In 1992–1993, my wife and I returned to Tanzania to revisit the 18 villages and reinterview as many as we could of those we had interviewed 21 years earlier. We asked some of the same questions again and tried to determine how farming had changed and why. We were able to interview 113 of those in the original survey, the spouses of 15, the children of 45, plus 9 other relatives. In all, we did 182 interviews, a rate of 78 percent of the original survey (not counting one village in western Tanga Region that we were not permitted to visit). (The 18 villages in our survey, incidentally, are as wonderfully different as would be 18 people chosen at random for interviews, say, on a New York City sidewalk.)

Farmers do not till their fields in a vacuum. Tanga has a rich and complex history, and to understand what farmers do, one must also understand the historical events and social and economic forces farmers have experienced over many years. Therefore, I did extensive research in the Tanzanian National Archives and the archives in the town of Tanga.[3]

The present study gives special attention to the environmental and social history of Tanga as a region, beginning with an analysis of the relations of Tanga's people with Zanzibaris in the 19th century (the slave and ivory trade, Zanzibar's clove industry, and Omani Arab sugar plantations on the Tanga coast, which had a labor system based on chattel slavery). I investigate the consequences of German colonial conquest (1889), the rinderpest pandemic (1891), the growth of the sisal industry (from 1892), and economic developments and policies under British colonial rule and, after 1961, under the Tanganyika African National Union (TANU). The nature of peasant society among Zigua, Sambaa, Nguu, Digo, Segeju, Akamba, and other cultural groups is considered with particular reference to political and economic events that affected the lives of Tanga's people. Changes in such things as clientship and dependencies are of particular importance, as is the story of how governmental policies (German and British) caused the local people to lose control of their environment, as livestock losses, disease, bush encroachment, and tsetse fly infestation increased. The story does not stop with Tanganyika's independence (1961). I also describe later developments, such as the consequences of *ujamaa vijijini* (familyhood, or socialism, in the villages), a system described by President Julius Nyerere in 1967 and implemented from about 1970 onward; the collapse of the East African Community and subsequent war with the forces of Uganda's Idi Amin; and the International Monetary Fund's structural adjustment program, which formally began in 1986.

My aim is to understand local farming knowledge and practice in two contexts, one general and one specific. The general context is farmer knowledge and practice as affected by the historical structures and events of the past 150 years, essentially environmental history and political economy. The specific context is farmer knowledge and practice in relation to water for crops. Farmers choose *where* to plant, *what* to plant, *when* to plant, *whether* to intermix crops in the field, *how* to space the seeds or cuttings they set out, and *whether* to thin or to do things to increase moisture to their plants or reduce the plants' need for moisture.

All of the just-listed aspects of farming are factual matters that surround the life of a plant in a field. They can be known and they can be modeled. In agroscience such models are called crop-water simulation models (Kirda 1999). The set of ideas against which I examine local knowledge concerns the agronomic requirements of crops. The plant lives or dies in relation to the flux of energy and moisture flows in the atmosphere, soil, and plant tissues. I have spent a long time refining a computer-driven model of energy-water-crop relationships that permits one to simulate crop performance, specifying all of the above-listed items a farmer must decide upon. In a simulation, one can hold all but one item—say, delay in planting—constant to see what the consequences on yield are if the farmer plants on time, is 14 days late, is 28 days late, etc. A portion of the study examines the above list of farmer decisions and then examines what farmers actually do in their fields with respect to the same list of decisions.

The simulations, although they produce yield numbers, are only estimates. Rather than exact predictions, they are suggestive of yield tendencies. One of the obvious limitations of crop-water models is that crop nutrition (fertilizers, mulches, etc.) is generally not included. Since most research in soil science is directed toward aspects of crop nutrition, I feel justified in limiting my analysis to crop and soil moisture (see chapter 6 and Appendix A).

Another theme is environmental and social-economic causes of food shortages and how people cope when there is drought and/or a food shortage. Coping occurs at different levels. Governments must cope, and they institutionalize policies and procedures regarding food shortages. The rural people—men, women, and children—also must cope. The famous Handeni famine of 1953 is considered at some length in chapter 7.

Tanga's farmers have considerable knowledge about the soils, terrain, and weather they encounter in their livelihood. They make many good decisions. Yet it was clear to us in 1993 that agriculture had not progressed very much for the majority of lowland Tanga's people since 1972, and even since 1961. Indeed, between 1972 and 1993 there was a troubling shift in crops being grown—an increase in maize and cassava and a significant decrease in vegetables, oilseeds (simsim [sesame], sunflower), and pulses (beans, peas, cowpeas, green grams [mung beans], etc.) (see chapter 7). This may indicate a potentially serious impoverishment in the local diet.

There is persistent continuing poverty in Tanga Region. A detailed inventory of material wealth done in the 18 villages in 1972 and again in 1993 showed that there has been virtually no change for the better. The same can be said of health. Population growth exacerbates the problems. In 1948 there were 462,000 people in Tanga Region (excluding Pare District); in 1957 there were 512,000; in 1967 there were 771,000; and in 1988 there were 1,280,000, an increase of 66 percent. Tanga Region's population grew to 1.6 million in 2002 (an increase of 28 percent from 1988). These figures correspond to annual rates of population increase of 1.2, 4.2, 2.5, and 1.8 percent. How life and livelihood can be improved in the face of these continuing increases has not been figured out.

I conclude that prior to the middle of the 19th century, Tanga's lowland farmers had worked out an economy that was generally in balance with nature, although not without serious food shortages and even famines from time to time (Giblin 1992: 8, 19ff.). The population was substantial but far less than it had become by the time of the first real census, which took place in 1948. The economy was based on farming small grains (sorghum, millets, rice, and maize), stock keeping, clientship, exchange at local markets, and control of regional trade. There was extensive trading (honey, iron, tobacco, ghee, livestock, medicines, "the most beautiful ivory, and the best grain, resins, wax, simsim and millet"), which linked the Wazigua with the town of Pangani on the coast and with the Maasai Steppe (quotation from Guillain 1843: 520, translated in Giblin 1992: 23; Kimambo 1996: 92). From about 1850 onward all this changed, most dramatically after 1890. Since then the local agricultural economy has never been stable and it has been difficult for farmers to improve their livelihood. One reason for the lack of economic betterment in Tanga Region is the relative isolation and neglect rural Tanga experienced under its succession of administrations (German,

British, and Tanzanian). Another reason is ill-considered interventions by these administrations. Yet another reason is Tanga's difficult environment, coupled with historic shifts in crop mixes toward maize and cassava and away from more drought resistant sorghums and millets. Finally, another contributing factor is the disconnect between institutional Western science and local knowledge and local society.

This last reason is exemplified in the relationship between local farmers and the staffs of the Mlingano Agricultural Research Institute (MARI) and the Ministry of Agriculture Training Institute (MATI), both near the town of Muheza. MARI is responsible for the nation's soil surveys and for agricultural experiments on crops and fertilizers. It operates independently and unmindful of the knowledge about soils and crops that local farmers have. MATI is responsible for the diffusion of technological improvements in agriculture. It has a two-year diploma program as well as short courses for farmers, students, agricultural extension workers, and other professionals. There are practices and choices Western science can suggest that could help Tanga's farmers. There are things Tanga's farmers know about their environment that could help Western science. Those who *study* and those who *tend* growing plants could benefit by working together and learning from one another.

For Tanga's people to undergo economic and social development and a transition to a system that is both environmentally and structurally sustainable, many things need to be done. Such a transition will require great investment in infrastructure and in the people themselves—roads, clinics, schools, markets, agricultural machinery, cooperative societies, industrial employment opportunities, and the like. Nonetheless, among fruitful avenues of approach are the exchanges that could be developed between Tanga's local farmers and research scientists and extension workers at MARI and MATI. Technical communication and technology transfer, modification, and adoption have a useful role to play.

The present study is relevant to the general problem of agricultural development in Africa. A recent United Nations Inter-academy Council report, *Realizing the Promise and Potential of African Agriculture*, describes a series of unique features that make Africa different from Asia, where the green revolution has had such a notable impact (Inter-academy Council 2004). Characteristics of African agriculture include (1) lack of a dominant farming system on which food security largely depends, (2) predominance of rain-fed agriculture as opposed to irrigated agriculture, (3) heterogeneity and diversity of farming systems and importance of livestock, (4) key roles of women in agriculture and in ensuring household food security,

(5) lack of functioning competitive markets, (6) underinvestment in agricultural research and development and infrastructure, (7) dominance of weathered soils of poor inherent fertility, (8) lack of conducive economic and politically enabling environments, (9) large and growing impact of human health on agriculture, (10) low and stagnant labor productivity and minimal mechanization, and (11) predominance of customary land tenure (Inter-academy Council 2004: xviii). All of these topics are considered in the present work.

Conceptual Contexts

This study contributes to several research themes: cultural and political ecology, crop-water modeling (a part of agroscience), risk-hazard vulnerability analysis, and sustainability science.[4] Each of these themes has an extensive literature, some of which is commented upon below. The literature on crop-water modeling is discussed in chapter 6. Each of these themes is part of an ever-changing mixture of human-environment research cores in geography that swing and spiral between natural science and humanistic perspectives (Turner 1997: 200). I see two main contributions of the present work. First, studies of risk and hazard management in less developed countries, though excellent in their analysis, vary widely in the degree to which they are empirical, quantitative, and grounded in the biophysical world. Commonly, their explanations are qualitative (Blaikie 1985; Blaikie and Brookfield 1987; Richards 1985, 1986; Chambers 1974; Mortimore 1998, 1999 [like this, a longitudinal study]; Mortimore and Adams 1999; Turner et al. 2003a; Turner et al. 2003b).

Chambers and Richards in particular adopt a populist, "farmer-first" argument, which asserts that agricultural institutions (experiment stations, extension services, and the like) should "build on the needs, ideas and knowledge of the rural poor" (Bebbington 1994: 206). Among other things, these authors describe what people believe and what they do to reduce risks in their livelihood, and the authors do it exceedingly well.

This study has similar goals, but the focus is narrowed to questions about farmers' management of crops in relation to the crop's need for water. Its main contribution is its rigorous, biophysical, quantitative approach to the matter of crop water, achieved through simulations. The simulations enable us to ask questions about the livelihood consequences that ensue when farmers make decisions about what to plant, where to plant, when to plant, and how to space their crops. I also inquire as to what other management practices they can engage in that help meet crop water needs or reduce the need for moisture for the crop. Through much experimental effort, agronomists and other scientists have been able to describe the workings of crops and the environments that surround them in quantitative biophysical terms. My simulations extend the analysis to include farmer decision making, with outcomes expressed as estimated yields flowing from the choices a farmer makes.

This study also differs from many ecological investigations in that it concerns a 250-km-long transect rather than a single, intensively studied community. It does not present a detailed case study, and it therefore lacks the kind of insights gained, for example, from Paul Richards's (1986) wonderful case study of rice cultivation in Mogubuma, Sierra Leone, or from Michael Mortimore's (1999) study of the close-settled zone north of Kano, Nigeria. The second contribution of the present work flows from the geographer's eternal task: description and analysis of the earth as the home of humankind. Sometimes geographers lose sight of this nominal task. Looking back on a career, a geographer can ask: How much of the earth's surface have I described and interpreted? Tanga Region is an area about the size of Maryland (26,548 km², or 10,248 mi²). It is a worthy task to present a regional geography of Tanga Region that includes environmental, historical, demographic, political, economic, and social analysis. Chapters 3 and 4 provide a regional synthesis as a context in which to understand the 18 villages in our rainfall transect and the actions farmers take to cope with the place where they live. Chapter 4, in particular, combines environmental history with political ecology.

Cultural and Political Ecology

The research contexts of this study include cultural and political ecology. Cultural ecology's central interest is in the mutual relations of people and the environment people use. The term "cultural ecology" was coined by Julian Steward in 1937 to distinguish it from biological and social concepts of ecology (Steward 1968). It concerns the cultural and technical processes by which a society adapts to its environment (Batterbury and Forsyth 1999). The history of cultural ecology has been traced many times (Vayda and McCay 1975; Grossman 1977; Porter 1978, 2001; Butzer 1989; Turner 1989, 2002; Bassett and Zimmerer 2003).

Most cultural ecology is concerned with a closed system in which human-environment relationships are examined. There are two general types: analysis of functioning systems and analysis of indigenous decision making. One emphasizes structure; the other emphasizes agency. Risk-hazard vulnerability and environment-development themes are prominent in such studies (Brokensha, Warren, and Werner 1977; Kirio and Juma 1989; Little, Horowitz, and Nyerges 1987; Wisner 1977; Moock and Rhoades 1992).

An important issue in cultural ecology concerns the question of who defines the terms of the research, and the answer has led to the development of a subfield in cultural ecology. My own personal preference has been to let the people being studied lead the way, and there have been others who have championed this approach, which relies on the internal view that people have of their environment (Porter 1965; Frake 1962; Conklin 1957). The approach involves the emic world of contextualized meanings of words, in contrast to any universal etic or standardized way of saying a word (terms derived from linguistics—phon*emic* and phon*etic*) (Pike 1966). It is, after all, local understandings and categories that inform what people know and do. This is their indigenous science, their local knowledge.

Political ecology is a more recent theme and represents a critical appraisal of some of cultural ecology's common shortcomings, as well as a critique of natural-hazards research (Blaikie and Brookfield 1987). The shortcomings of cultural ecology can include (1) assuming equilibrium and balance in a society, (2) using an ethnographic present (as if the people studied were unchanging and had no history), (3) using a kind of circular reasoning in presenting a functionalist argument, and (4) failing to incorporate influences from outside the group (Zimmerer 1994). The books that brought political ecology to the forefront were *Silent Violence* (1983) by Michael Watts and *Land Degradation and Society* (1987) by Piers Blaikie and Harold Brookfield. *Silent Violence* showed how peasant farmers in northern Nigeria were ineluctably enmeshed in political and economic circuits that operated at national and global levels. Through political ecology, greater attention has come to be paid to the roles of the state, class, gender, ethnicity, social movements, and identity (Blaikie and Brookfield 1987; Carney 1993; Rocheleau and Edmunds 1997; Bebbington 1996; Forsyth 2003; Escobar, Rocheleau, and Kothari 2002; Moseley and Logan 2004; Zimmerer and Bassett 2003). Political ecology is not without its critics (Blaikie 1998; Vayda and Walters 1999; Forsyth 2003). Among criticisms of some work in political ecology is relative neglect of the biophysical environment, generalized list making (featuring + or − rankings), and "unspecified reference to victimization or oppression . . . which is one symptom of the 'academic hitchhiker's' view of political ecology" (Zimmerer and Bassett 2003: 276). Vayda and Walters (1999: 167) fault political ecology for missing "the complex and contingent interactions of factors whereby actual environmental changes are produced" and propose in its stead an approach they call event ecology.

Risk and Hazard Analysis

Risk and natural-hazard analysis has a history that parallels, but in many ways has been distinct from, cultural ecology. It started at the University of Chicago with Gilbert White and some of his students (Burton and Kates 1964). To begin with, it considered the perceptions people had of where they lived—flood-prone floodplains and coastal settlements subject to occasional damaging storms and beach erosion. The approach to natural hazards was subsequently applied globally to hurricanes, tropical cyclones, floods, coastal erosion, drought, freeze hazards, volcanic hazards, earthquakes, and avalanches (White 1974; Kasperson, Kasperson, and Turner 1995; Kasperson and Kasperson 2001). The approach has also been applied to hunger, warfare, and violent environments (e.g., enclosures of genetically engineered crops, nuclear and militarized zones, and industrial-disaster settings, such as Bhopal, India)—though they are clearly not *natural* hazards (Hewitt 1987; Peluso and Watts 2001; Turner et al. 2003a). Among criticisms of risk-hazard analysis has been its use of standardized questionnaires across cultures, which tended to ignore local contexts (Waddell 1977).

Risk-hazard research has evolved from its roots at the University of Chicago into vulnerability analysis (Blaikie et al. 1994; Turner et al. 2003a). A vast literature has grown out of this development, reporting research conducted at global-international, regional, and local scales—from climate change and global environmental change to local, place-based study (SCOPE 1978; Cash et al. 2003). Place-based research opens the possibility for "integrating lay and expert knowledge" (Turner et al. 2003a: 8076). Researchers in vulnerability analysis seek to go beyond two standard models, RH (risk-hazard) and PAR (pressure and release), to a multiscale, coupled-system vulnerability framework.[5]

Sustainability Science

Sustainability science emerged out of several preoccupations: sustainable development, globalization, and global environmental change (Kates 1995;

Kates et al. 2001). "Sustainable development" became a popular term in the 1980s and 1990s. The term "sustainability," however, was so broad and so variously defined as to be operationally meaningless. Williams and Millington (2004: 99) cite a study that gives "80 different, often competing and sometimes contradictory, definitions." Ruttan (1994: 211) notes: "It is not uncommon for a social movement to achieve status as an ideology while still in search of a methodology or a technology." At the core of the idea of sustainability, however, is the use of natural resources in such a manner that the stock is renewed, kept in balance, and not depleted so as to make it unavailable to future generations (Bruntland 1987; Pearce, Barbier, and Markandya 1990). The concept continues to inform much research (Fernando 2003; Müller 2003).

Sustainability science is best thought of as one of "multiple sciences addressing a common theme" (Clark and Dickson 2003: 8059). It incorporates global and global-change perspectives and includes an approach called vulnerability analysis (discussed above)—that is, "the capacity to treat coupled human-environment systems and those linkages within and without the systems that affect their vulnerability" (Turner et al. 2003a: 8074; Turner et al. 2003b; Wilbanks and Kates 1999). The concept of a transition toward sustainability is a key feature of this research (Kates 2000). The purpose of a sustainability transition is to stabilize "future world population while reducing hunger and poverty and maintaining the planet's life-support systems" (Parris and Kates 2003: 8068). It appears easier for organizations to set goals and targets for reducing hunger and poverty than for maintaining the environment, or "life-support systems" (Parris and Kates 2003). In setting goals, identifying quantitative targets and associated indicators, formulating plans of action, and seeking internationally negotiated agreements, sustainability science makes a conscious search for the methods and techniques Ruttan called for.

————

The present study relates to every one of the above-discussed themes. The regional geography of Tanga (chapter 4) traces ways in which political and economic forces have affected the people of the 18 villages we studied—the political ecology theme. Throughout the study I pay attention to what people believe about the environment they use, how they use it, and with what result—the cultural ecology theme (chapters 2, 3, and 6). I also examine what people do when, despite their efforts, they fail to make a livelihood through farming—more cultural ecology as well as risk-hazard vulnerability analysis (chapter 7). Sustainability science is addressed in chapter 8, where I compare the success story of Machakos District, Kenya, with what has occurred and what may yet occur in Tanga Region. One feature of the Tanga story is the disjuncture between local knowledge and Western science knowledge as exemplified by MARI and MATI (chapter 8). Tanga Region, like many parts of the less developed world, faces the difficult task of improving the livelihoods of local people while the population constantly increases. How intensification and greater productivity can be brought about in tandem with increasing population pressure on local resources is a difficult puzzle to solve (Turner, Hyden, and Kates 1993; Turner and Brush 1987). Yet if local knowledge and that of Western scientific institutions could be brought together, with respectful participation on both sides—granted, no easy task—the result might be improved food security and human well-being.

Cultural ecology, political ecology, and risk-hazard vulnerability analysis have all followed divergent paths in recent years, and it is not easy to characterize trends, because one can find fissiparous tendencies within each tradition and common features that link across them. One cluster seeks to relate to "normal science" and features land-use/land-cover change analysis. It uses remote sensing, satellite imagery, and GIScience (Geographic Information Science) (Walsh, Evans, and Turner 2004; Taylor 2004). Another group challenges "normal science" as a way of knowing and uses social theory and critical theory to address questions of access, entitlement, and empowerment among marginalized people. This group favors structure over agency and may make only minimal efforts to integrate biophysical aspects into their analysis (Blaikie et al. 1994). It may reject received wisdom and environmental orthodoxies (such as those used to explain desertification, environmental degradation, and deforestation) and develop counternarratives, including those generated by indigenous voices (Forsyth 2003; Zimmerer and Bassett 2003).

Sustainability science attempts to reconcile the differences in the various approaches, but it is not easy given the different objectives, epistemologies, ideologies, and methods involved in each tradition. I hope this book demonstrates that there is much to be gained by respecting and drawing upon the various framings, methods, and objectives of these research perspectives. I have tried to do so in this work, although I must acknowledge that my original research question (What do farmers understand about the environment they use?) remains for me the most important one.

CHAPTER TWO

Livelihood and Poverty in Tanga Region

Peasant Farming in Lowland Tanga

Let one farmer represent many. Hamisi Rajabu (the name has been changed) was born in the town of Panga in Handeni District (Diboma Parish/Division). At the time of our visit, in 1972, he is 46 years old and has lived in Mgera for about 30 years. He has two wives, aged 30 and 22, a daughter and two sons by his first wife, and a daughter by his second wife. The eldest child is 10. Rajabu's most profitable cash crops in order of importance are tobacco, beans, and castor seeds. His most important food crops, again in order of importance, are maize, cassava, and beans. He has about 50 orange trees, 6 lemon trees, and 5 tangerine trees.

At the time of independence (1961) his most important crops were maize, castor seeds, beans, and tobacco, with maize being the most important food crop, and the others important cash crops. So the crops that were important to the family in 1961 continued to be important in 1972. Rajabu has not stopped growing any of the crops he grew earlier, and there is no change in the source of his seed. The orange trees, however, are new since 1961.

The Agricultural Season

Although there is only one rainfall peak (April) in Mgera, Rajabu divides the year into two seasons, *vuli* and *mwaka* (*masika*). The *vuli* rains can begin in September or as late as December, while the *mwaka* (*masika*) rains can commence in December or be delayed until April. Rajabu depends on cassava, since he cannot be sure that his maize crop will succeed. Cowpeas and green grams (*choroko*, or mung beans) are planted in January, reach an unripe stage in March, and are picked and dried during April. Green grams are interplanted with cassava. Maize and castor are planted in February and the maize can be harvested in May. Beans are planted in April and May. Tobacco is planted in May.

The Farm and Its Operation

Rajabu uses different areas of his farm differently. He plants in both upland and lower areas so that if the upland crops fail because of drought, the crops in the valley will give a yield. Other reasons for using different areas are that doing so allows him to grow different crops, and land is readily available.

Cassava grows well in the uplands, while beans, tobacco, and maize grow on both valley and upland areas. The worst year for crops in his experience was 1970, when the rains failed. There was famine, but people and stock did not die and water could be found. To cope, Rajabu took wage employment from a neighbor. One cannot influence the rains, and without irrigation, there is nothing one can do to help crops get the moisture they need to mature.

Rajabu has five fields in cultivation and no other fields. He is not using anyone else's land. His fields are not contiguous. He just took the land he needed and did not get it from his father or another relative. One of his wives, who is not living with him, also uses the land. There is still plenty of land in this area for the taking. His children will inherit his land. It is about 1 mile from his home to his nearest field, which he can reach in about half an hour, and 2 miles (3 km) to the most distant field. Land disputes are not a problem in this area. Rajabu and his family do not collect honey or try to catch fish. In the dry season his wives and children collect a tuber (*ndiga, Dioscora dumetorum*) and wild vegetables. He has no other income-producing work, aside from cutting firewood. He and other family members participate in work parties for bush clearing, cultivating, weeding, and harvesting. They get water from the well in all seasons. In years when the rains are good, they get water both from the well and from the river. In a bad year, they get water from the well only. It takes half an hour to walk to the well. His wives get water twice a day and carry it back in pots and buckets on their heads. Rajabu does not buy water. He uses about 28 *debes* (1 *debe* = 4 imperial gallons, or 18.18 liters) of water a week, storing it in pots and pails. The family does not boil their drinking water. In the dry season the water becomes salty, but it creates no health problems. His family of six, with four youngsters 10 or younger, uses about 12.1 liters/person/day. This compares with rates of 7.8 and 12.7 liters/person/day in two rural villages in Tanzania (Mkuu and Kipanga) in about 1969, and 14.2 and 16.6 liters/person/day in the same villages in the late 1990s (White, Bradley, and White 1972: 119; Thompson et al. 2004: 110).

Rajabu does not buy seed. He exchanges stores of food with relatives and neighbors. He harvested maize, cassava, and beans (four *debes*) this past season. He sold two *debes* of beans in sacks. He has not sold maize or cassava. The maize is piled in the cob in his house, about six *debes*, and the cassava is left in the field. He has problems with rats and insects in his stored food and estimates his losses as being one *gunia* (a 100-lb. bag) a year. To protect his beans, Rajabu uses *dawa* ("medicine"), a powdery insecticide with the local name of *dipu*. He buys it from a shopkeeper. It costs T Sh 2/- (2 Tanzanian shillings) per handful and gives good results, and he plans to continue using it. There is also trouble in this area with thieves stealing food.

Rajabu's family usually pound their own grain, but sometimes they take it to the mill. In a year of good harvest, he will sell some of it. He also sells beans, tobacco, and castor seeds, often to the cooperative society. He sells beans to the cooperative society at the market every year during the dry season. He sold a small quantity of tobacco once (1970) to local traders who came to his home.

Management of the Most Important Crops

Maize and beans are planted at different times, but in the same field and during the same rainy season. Rajabu mixes crops because less labor is thus required for weeding. Interplanting also uses rainfall effectively and decreases the risk of crop failure because with many crops some crops may succeed even if others fail.

The crops Rajabu grows in pure stands are tobacco and sometimes cassava. He does not prepare a field at the beginning of the *vuli* rains for planting at the beginning of the *mwaka* rains. According to Rajabu, that is not a good idea. He does not use farmyard manure or commercial fertilizer. He grazes his goats in the field after harvest but does not allow others to graze livestock there. Tools for farming present no problem; he can buy them when he needs them, and the cost is not too high.

He uses local varieties of maize although he has heard about other varieties. He knows, for example, that Katumani matures earlier and is faster growing by about 30 days. He has seen it. In pest and disease resistance, the local maize varieties are the same as Katumani maize, and they are as easy to shell, have the same taste, and require less labor. He says he plants maize in an area of about 3 acres (1.2 ha).

In planting Rajabu follows a crop calendar. Maize is planted one day after the beginning of the rains, both in *vuli* and in *mwaka*. He is not aware of the District Crop Calendar (a set of recommended times for planting, weeding, and harvesting different crops grown in Handeni District). He plants at a rate of one and one-half a *debe* of bean seeds/acre (75 kg/ha) and one tin of maize/acre. If the rains have a rapid onset, he plants less cassava. He is increasing the amount of land he is cultivating and therefore the amount of seed he plants. He designates some of his maize harvest solely for domestic consumption, and he always sells some of his bean crop.

Both he and his wives do the work of clearing. They do not clear new land every year. They use panga (machete), hoe, and axe. The amount of time an individual devotes per day to agricultural tasks during the agricultural season (October through August) is given in table 2.1.

There are labor bottlenecks in Rajabu's farming, during clearing and digging. It is common for people in this area to employ others (*kibarua*) to help with the farming. They help with clearing new fields, digging, and weeding but not with planting and harvesting. Rajabu mentioned that he had employed people this past year for two weeks to help cut down the

TABLE 2.1. Labor Inputs by Family of Hamisi Rajabu

Input	Hours/day	No. days	Total hours
Clearing time	6	30	180
Digging	12	14	168
Planting	12	1 or 2	24
Weeding	6	14	84
Guarding	12 (the maize)	144	1,728
Harvesting	8	7	72
Postharvest (to increase the size of the field on its edges)			224
Other (work parties)			192
Total (at least)			2,672

bush in his maize field. He employed some others to clear land for his bean field. He could not recall the number of people he employed.

He was ill this year for one week from diarrhea and vomiting. His first wife was ill with diarrhea and vomiting for two weeks.

I estimate that Rajabu and his family and occasional employees devoted at least 2,672 hours to the fields in the five-month period from January to May when fields are prepared and crops tended. This works out to about 4.5 hours/day/person, although the labor demands are concentrated, with 12-hour workdays at the beginning of the season. It also does not count the labor women devote to other tasks, such as cooking, sweeping, carrying water, collecting and carrying firewood, caring for children, etc. These figures agree generally with those presented by Sumra (1975a: 127).

Mapped Fields

We mapped two fields. One field (a small one, about 0.4 ha) is planted in cassava and Rajabu takes care of it. His family controls the harvest from it. It was not interplanted with anything else. It lies partly in *mbuga* (a low, flat area, although this one never floods) and partly in upland. The steepest slope observed was 9 degrees. The soil type is red (*udong mwekundu*). It has been used for cassava for the past three years and was in bush before then.

The second field we mapped belongs to his wives, and the whole family controls the harvest. It is mainly interplanted with orange trees, beans, and maize and castor plants. The area is upland red soils. The field has been in cultivation for the past five years: first in maize for one year; then maize, beans, and castor for two years; and finally the same crops plus orange trees in the two most recent years. The field is rectangular and measured just under 2.5 ha. The crop triangle samples we took suggest that the field contains over 100 orange trees and 12,000 castor plants. The maize had already been harvested.

———

This summarizes part of what we learned about Hamisi Rajabu's farming in Mgera in 1972. In 1993, he was 67 years old; we found him still active and living in Mgera, although slowed down somewhat by age. Since 1972 he had stopped growing sorghum (a crop that he did not mention to us in 1972) and was planting less maize, because he had less strength for farming. He noted a decline in yield of all his crops but said that this was because he and his wives had less energy to devote to them. His children were in their

TABLE 2.2. Time Taken to Go for Water and Fuel Wood

Village	Water 1972 (minutes)	Water 1993 (minutes)	Fuel wood 1993 (hours:minutes)
Kwadundwa	13	—	—
Mgera	12	12	2:04
Kwediamba	45	22	1:34
Minazini	61	63	0:53
Kwamsisi	52	36	1:03
Mzundu	46	39	0:42
Kwamgwe	9	36[a]	1:42
Mandera	14	61[a]	1:51
Mlembule	8	13	0:53
Mkomazi	12	12	4:26
Vugiri	13	8	0:42
Magoma	69	30	3:57
Kisiwani	9	6	1:13
Kiwanda	11	11	0:26
Daluni	14	5	2:53
Mwakijembe	28	8	1:53
Maranzara	20	39[a]	2:39
Moa	9	23	1:33
Average	24	26	1:44

[a] In these three instances, a source (either a well or a borehole) that had been used in 1972 was no longer usable.

late twenties and early thirties and now living elsewhere. Hamisi Rajabu did not represent any special type of farmer in the sample (old, young, rich, progressive); he was just an average farmer in Mgera.

Water and Fuel Wood

We asked both in 1972 and in 1993 how long it took to go for water. In 1972 we also asked who went to get it, and how many times a day they fetched water. Almost universally, it was women, aided by children, who fetched water, usually two to three times each day. The amount of time taken varied greatly between villages and, in some instances, between residents of the village, but overall, there was little change between the two dates (table 2.2). The collective time devoted to obtaining water (assuming the same rates of visits to the water sources in both periods) was about 24 minutes in 1972 and 26 minutes in 1993, virtually no change.

Thompson and his colleagues in their follow-up study of Gilbert White, David Bradley, and Anne White's classic study of water use in East Africa (*Drawers of Water*) found that the average time it took people in two rural areas in Tanzania to fetch water (late-1990s) was 40.2 minutes/trip and the average number of trips per day was 4.0 (White, Bradley, and White 1972; Thompson et al. 2004: 60, 110). Some of the time was probably spent standing in line at the watering place. Thompson and his coworkers found that with a greatly increased population (growing in three decades from 32 million to over 83 million people) congestion and queuing times had greatly increased at water points (Thompson et al. 2004: 97). Service disruption of piped facilities was greatly increased, reducing the hours when water was available (Thompson et al. 2004: 99). What we found with respect to changes in water supply in the 18 villages between 1972 and 1993 reflects the finding of the two larger *Drawers of Water* studies.

An explanatory note is in order at this point. The order of villages given in table 2.2 places the villages in a spatial sequence going from southwest to northeast (see fig. 1.2). This is the order used in all tables listing the villages. (This is not the order in which we visited villages and conducted interviews and field mapping. When we returned 21 years later, we did our work in the villages in a totally different order, since we were working out of Muheza rather than Handeni. Finding these villages again, and the people whom we had interviewed in 1972, was an adventure with numerous surprises and false leads.)

It is much more time-consuming to walk to and cut fuel wood than it is to fetch water, although this task is not done several times a day. In a few places, like arid Mkomazi and long-settled areas like Maranzara and Magoma (where sisal estates dominate the landscape), one must go a long way to find a place where wood can be cut or collected. On average, households in the 17 villages spent an hour and three quarters on a quest for fuel wood. The high figure for Daluni is somewhat surprising. One person said that he purchased fuel wood at the market (no long trip there), whereas another claimed that it took him six hours to obtain wood.

Land in Cultivation and Field Sizes

We mapped 201 fields belonging to the 241 farmers we interviewed. This represented a 39 percent sample of the 522 fields these farmers had under cultivation in 1972. They also mentioned a further 120 fields in bush, that is, not currently in use. Of course, in many parts of lowland Tanga there is much bush, so it would have been easy for most farmers to claim still more land for cultivation had the need arisen. The 201 mapped fields covered an area of 134.02 ha (1.34 km^2), and the average size was exactly two-thirds of a hectare (0.667 ha, or about 1.65 ac). The median size of field was 0.47 ha, and 80 percent of all fields were smaller than 1 ha (fig. 2.1).

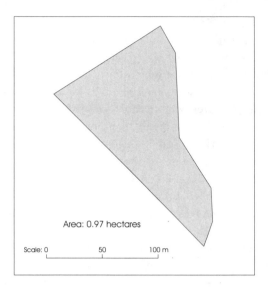

FIG. 2.1. Field of Zakaria Pilimila, Kwamgwe. Zakaria, his wife, and six children grew maize in this field in 1972. Other crops he grew were cassava, bananas, yams, sweet potatoes, and coconuts. He got the field from a relative.

Since we know the number of fields each farmer had, we can estimate farm size by multiplying the average field size by the number of cultivated fields. The result suggests that the average farmer had 1.45 ha under cultivation in 1972 and a further 0.33 ha in bush. There would be considerable variation in farm size, depending on family size, wealth, age, and other characteristics of the family. The Food Studies Group's study of six agro-ecological regions in Tanzania showed that relatively few farmers had land holdings in excess of 4 or 5 ha. For example, in Newala (a cassava and/or cashew and coconut area similar in many ways to lowland Tanga), only 11 percent of holdings were larger than 4 ha (Food Studies Group 1992b: 28). Holdings larger than 5 ha were respectively 1 percent in the banana/coffee and dairy system in Hai District, 2 percent in the agro-pastoralist system in Dodoma Rural District, and 9 percent in the livestock/sorghum-millet system in Kwimba District (Food Studies Group 1992b: 144, 124, 103).

In our own data we searched for a relationship between field size and number of fields and found $r^2 = 0.009$ (i.e., the relationship explains less than 1 percent of the variance), and thus there is absolutely no relation between field size and the number of fields a farmer has. One cannot determine farm size by extrapolating from the sizes of the fields in them.

Livestock

In a sample of 239 households, only 83 owned livestock. Forty-four households (18 percent) owned cattle, and the total herd of lowland Tanga, among the sample, was just over 400 head. The modal herd size was between one and five animals. During the rains all these herds together had about 120 animals providing milk, and if one divides 120 milking cows among a resident population numbering 1,494, including all household members, it is clear that milk and other dairy products are not important sources of food for the population. Six villages (Mlembule, Vugiri, Mandera, Minazini, Kwediamba, and Mwakijembe) where stock are present in substantial numbers are an exception. Goats figured importantly, with 62 farmers (26 percent) owning about 400 goats. There were 28 farms with sheep, numbering only 120. Goats and sheep were mainly eaten, not milked.

Over 80 percent of cattle are found in six villages, five of them in the same part of Tanga Region—central Handeni and central Korogwe Districts. This territory lies along historic stock routes through Handeni and lowland Korogwe. In western Handeni District lies a zone of tsetse fly infestation. Fly is also found along a coastal belt in Pangani and Tanga Districts. One outlier of cattle keeping is Mwakijembe, whose residents (Kikamba and Taita speakers) have a long tradition of cattle keeping.

In areas where both cattle and small stock were kept, provision of pasture is the overriding concern of most farmers, with livestock diseases (including trypanosomiasis) being cited as a serious problem as well. In places where cattle are important and goats and sheep largely absent (Moa and Mandera), people said that there were few problems associated with stock raising, citing only low sale prices and lack of medicines and veterinary services. Where cattle were absent but goats were kept (Mgera, Kwadundwa, Magoma, Mkomazi, Daluni, and Kisiwani), about half the owners of goats said either there were no problems or there were some relating to diseases, including foot rot. Five farmers cited wild animals as being problems. Thieves and water for stock were also problems mentioned by a few people. The pattern of stock ownership reflects the map of tsetse fly infestation and the terrain of Tanga Region (see chapter 4). The places where cattle are absent correspond to the fly zones, and where cattle and small stock are both few in numbers one finds forested hill country (Kiwanda, Kisiwani, and to an extent Daluni), on the east-facing slopes of the eastern Usambara Mountains. In general, owners of cattle have more land and are in a position to survive droughts and food shortages more readily (Sumra 1975a: 172). Overall, however, livestock played a minor, subsidiary role in the economy of the farmers of lowland Tanga in 1972 and 1993.

Poverty

Poverty is difficult to define and difficult to measure. Adolfo Mascarenhas did a study in Lushoto District that included both highlands and lowlands (Mascarenhas 2000). The six areas he studied included Bumbuli, which lies 23 km due north of Vugiri; two lowland plains villages on the Umba Plains to the northwest (Mnazi) and the northeast (Mlingano); two inland valley sites (Mbelei and Mlola) in the high central Usambaras; and lastly a village in the western part of the district, perched on the edge of a steep escarpment (Mtae). (In many ways, the sites complement the 18 villages we examine in this study.) He estimated that fully 50 percent of the population had incomes below the poverty line (Mascarenhas 2000: 35). Income alone is not a sufficient measure of poverty. Sarris and Brink, using income and diet, estimated that in 1976–1977, 67 percent of Tanzanians lived below the poverty line (Sarris and Brink 1991: 211). Mascarenhas used several other measures of well-being: a "wealth index" (based on an approach of Fleuret [1978, 1980]), an assessment of "disadvantages leading to poverty" (this is a cluster of five self-perceived conditions devised by Robert Chambers [1983: 108ff.]), and an assessment of food security in the household. Table 2.3 summarizes the main results of these analyses (Mascarenhas 2000: 31–35).

The 154 people Mascarenhas interviewed attributed problems related to their well-being as follows: powerlessness, 45 percent; vulnerability, 42 percent; physical weakness, 38 percent; isolation, 19 percent; and poverty itself, 15 percent. Powerlessness and isolation were greatest among plains pastoralists and plains agriculturalists; physical weakness and vulnerability were high among plains agriculturalists and poor highland agriculturalists.

Mascarenhas's goal was to understand links between environment and poverty. It is not surprising that he found no direct links, even though there are plenty of indirect ones. There are social and economic factors as well. The poorest tend to be pastoralists and self-provisioning, female heads of households. Commercial farmers in highland areas exhibited the lowest level of poverty. The dream of most for themselves was to practice

TABLE 2.3. Dimensions of Poverty in Lushoto District, Tanga Region

	Plains pastoralists (%)	Plains agriculturalists (%)	Mountain poor (%)	Mountain rich (valley) (%)
Income[a]				
<100	37.5	52.2	41.9[b]	44.0
100–250	37.5	8.7	35.3	3.0
>250	12.5	17.4	20.4	53.0
No data	12.5	21.7	—	—
Wealth index[c]				
Destitute	18.8	23.4	20.0	11.1[b]
Poor	31.3	34.5	26.7	33.3
Average	50.0	33.1	33.3	33.3
Above average	0	8.7?	20.0	16.1
Nonmaterial				
Powerlessness	43.8	62.0	29.4	23.5
Isolation	31.3	25.7	8.8	8.8
Poverty	25.0	20.0	14.7	0
Vulnerability	25.0	61.4	42.2	11.8
Physical weakness	56.3	47.1	32.3	14.7

Source: Mascarenhas 2000: 31–35.

Note: n = 154.

[a]In thousands of Tanzanian shillings.

[b]There is possibly error in the figures in this part of the column, since they do not sum to 100.

[c]The wealth index was based on 23 items: house structure, capital-heavy items (bicycle, watch, radio), agricultural tools, and house furniture and fixtures.

modern agriculture and possibly run a business; for their children (both girls and boys), it was for them to have good schooling (although there is still a strong hope that girls will marry and become good wives).

Returning to our study of 18 villages in Tanga Region in 1972 and 1993, what is the status of those we interviewed with respect to poverty, wealth, well-being, and hopes for the future? Although we do not reduce or combine the data presented in the remainder of this chapter to construct a wealth index or a table of "nonmaterial aspects leading to poverty," we do present in a qualitative way much of the material that would go into such indices.

Illness and Lost Workdays

Both in 1972 and 1993, we asked people questions about illness and the number of days they were unable to work. In 1972 the total number of people in the sample was 855. Of the 135 household heads (nearly all men) who gave information about illness, 129 cited the number of days they could not work. They lost a total of 10,217 days, on average 79 days each, or the equivalent of 28 person-years. The spouses, children, and other household members numbered 158. Of these, 143 gave information, indicating that they lost a total of 3,893 days, on average 27 days each, or the equivalent of nearly 11 person-years. Eight respondents and 13 others (nine of them wives) were incapacitated for an entire year. For example, a blind person cannot work in fields and will likely contribute little to the year's labors. The same may be said of a very old person. If 855 people lived out a full year and the equivalent of 39 person-years was lost, this represents a 4.6 percent loss of potential labor, leaving aside the fact that not all children are old enough to work. Indeed, most children were ill only for a few days, up to a couple of weeks. It was the adult men and women who listed periods of a month or more when they were not able to work.

Given the way the data were gathered (recollections by those whom we interviewed), we should take away only a general impression of the sorts of illness that prevented people from working (table 2.4). The most frequently mentioned were malaria/fever and intestinal problems (40 percent), followed by what was commonly referred to as "body weakness" (10 percent). Headaches and respiratory problems were also mentioned often (14 percent). Eleven women lost some time because of pregnancy or childbirth.

As to the 1993 data, a problem in the questionnaire (the many possible meanings of the preposition *katika,* or "in") generated information about past years, not simply "this past year," and thus comparable results were not obtained. People were remembering back to 1991, 1985, 1978, etc. An inspection of the answers, however, shows the same mix of complaints, and the distribution and range in number of days are similar. It appears that levels of health among the people are neither dramatically better nor worse than in 1972.

In no instance did anyone interviewed in 1993 mention acquired immune deficiency syndrome (AIDS) or human immunodeficiency virus (HIV), things that did not exist in Tanga Region in 1972. In the small hotels where we stayed, however, free condoms were on offer. For example, at the Memnango Guest House, Maramba, condoms were available in a dish at the reception desk, and above the beds in the rooms were garish posters featuring a man and a woman in bed, with warnings about the risks of having unprotected sex. Other evidence of the presence of AIDS had to do with the Anglican-run hospital in Muheza. Funding in 1993 was so short that they had to operate reusing rubber gloves, and at times operations

TABLE 2.4. Causes of Lost Work Days, 1972

Condition	Men	Women	Children
Malaria, fever	37	14	25
Intestinal, ulcer, stomachache	23	26	6
Muscular, general body weakness	13	16	2
Pregnancy	—	11	—
Headache	9	9	7
Respiratory, cough, flu	11	7	2
Heart troubles	—	6	—
Eye problems, blindness	2	3	—
Pneumonia	5	3	1
Toothache	—	7	—
Hookworm, filariasis	2	2	3
Vomiting, nausea	3	5	2
Dysentery, diarrhea	4	3	2
Measles	—	—	1
Bilharzia	2	—	—
Urinary	1	—	—
Asthma, allergy	3	—	—
Polio	—	—	2
Anemia	1	1	—
Chest problems	5	2	—
Tuberculosis	3	1	—
Typhoid	1	—	1
Hepatitis	1	1	—
Leg problems	4	3	1
Gall bladder	—	—	2
Hernia	7	—	—
Injury	4	—	1
Other (skin, etc.)	3	1	2
Old age	2	—	—
Total	146	121	60

were canceled because of a lack of gloves. The doctors and nurses were at great risk because of the prevalence of HIV infection and AIDS (Porter and Porter 1993: 58). Aside from these examples, however, it was as if AIDS and HIV did not exist in Tanga Region, since no one mentioned them.

That AIDS has had a considerable impact in Tanga Region in recent times is suggested by data in the Tanzanian 2002 census (Government of Tanzania 2002). In Tanga Region for the cohort of men and women between the ages of 20 and 34, the ratio of women to men is 123 to 100. Stated another way, there were 46,056 fewer men than there were women (158,856 vs. 204,921). One would expect the numbers in that age cohort to be roughly the same. Looked at more closely, the pattern of difference in numbers of men and women is the same in rural areas as in urban areas.

TABLE 2.5. Infant Mortality Estimates, Tanga Region, 1967

District	Infant mortality[a]
Handeni	129
Korogwe	124
Lushoto	81
Pangani	103
Tanga	97
Tanga Region	106

Source: Thomas 1972.
[a] Deaths per 1,000 live births.

Using the 1967 census of Tanzania, Ian Thomas (1972) estimated infant mortality using questions the census had asked regarding the number of children ever born to a woman and the number of children born who were still surviving. The data for Tanga Region are given in table 2.5. The average for all of Tanzania is 139 deaths per 1,000 live births, so although Handeni and Korogwe Districts are high compared with other parts of Tanga Region, they are below average for Tanzania as a whole.

Infant mortality is frequently linked to dietary factors, diseases of early infancy, pneumonia and bronchitis, and gastroenteritis. Thomas did an analysis of possible conditions affecting infant mortality using a product moment correlation approach and found several variables to be statistically significant, although collectively they explained only 35 percent of the regional differences. The important factors were family size, women's education, gross domestic product, and number of livestock units (Thomas 1972).

Material Culture

A settled landscape contains clues about how well or poorly a group of people may be doing. Both in 1972 and in 1993, we asked questions about material aspects of the lives of those interviewed. If a person owns a bicycle, wears a wristwatch, and has a new *bati* (corrugated iron) roof on the family's house, it represents expenditures above and beyond subsistence. The answers to questions we asked about material goods (many of them helpful to livelihood) are summarized in tables 2.6 and 2.7.

The main purpose of examining the ownership of things (or, in one instance, access to a tractor) is to see whether the people we interviewed

TABLE 2.6. Material Objects Owned by Respondents, 1972

Village	Plow	Share tractor	Maize mill	Cart	Tilley lamp	Radio	Watch or clock	Bicycle	Pikipiki or truck	Bati roof	Cement floor	Brick house	Water catchment	Casement windows	Items/ person
Kwadundwa	0	0	0	0	0	2	2	4	0	1	0	0	0	0	0.73
Mgera	0	0	0	0	0	1	0	0	0	2	1	0	1	1	0.50
Kwediamba	0	1	0	0	0	1	1	3	0	1	0	0	1	0	0.73
Minazini	1	0	0	0	0	3	2	2	0	1	0	0	2	1	0.85
Kwamsisi	0	0	1	0	0	4	2	5	0	2	1	0	1	0	0.80
Mzundu	0	0	0	0	0	0	1	1	0	0	0	0	0	0	0.15
Kwamgwe	0	0	0	0	1	7	8	7	1	3	4	1	2	1	2.77
Mandera	0	2	0	0	0	6	3	6	0	3	2	0	0	1	1.60
Mlembule	1	8	0	0	0	2	3	8	0	1	0	0	0	0	1.53
Mkomazi	2	9	0	1	3	3	5	2	0	5	3	1	1	2	2.93
Vugiri	0	0	0	0	2	4	5	0	0	7	2	1	2	2	1.92
Magoma	0	0	0	0	0	3	0	3	0	1	0	0	1	0	0.47
Kisiwani	0	0	0	1	1	4	5	0	1[a]	5	2	0	2	4	1.51
Kiwanda	0	0	0	0	0	2	2	0	0	1	0	0	1	0	0.62
Daluni	0	0	0	0	3	5	5	1	0	4	4	0	2	0	1.50
Mwakijembe	0	0	0	0	0	1	3	2	1[a]	2	1	0	2	0	1.03
Maranzara	1	1	1	0	0	3	3	3	0	0	3	0	0	0	1.33
Moa	1	0	0	0	1	3	2	2	0	0	2	0	0	1	0.93
Total	6	21	2	2	11	54	52	49	3	39	25	3	20	12	299
No. of interviews	246	246	246	246	246	246	246	245	246	238	238	238	238	238	
Percentage	2.4	8.5	0.8	0.8	4.5	22.0	21.2	19.9	1.2	16.4	10.5	1.3	8.4	5.0	1.23

[a] A truck (gari). A pikipiki is a motorbike.

TABLE 2.7. Material Objects Owned by Respondents, 1993

Village	Plow	Share tractor	Maize mill	Cart	Wheelbarrow	Tilley lamp	Radio	Watch or clock	Pikipiki	Bicycle[a]	Bati roof	Cement floor	Brick house	Water catchment	Casement windows	Items/person
Mgera	0	0	2	0	0	1	3	0	0	1	2	1	0	0	0	1.36
Kwediamba	0	0	0	3	1	1	3	3	0	—	3	0	0	1	0	2.20
Minazini	0	0	0	0	0	0	5	6	0	5	3	1	0	0	0	1.67
Kwamsisi	0	0	0	0	0	0	6	4	0	—	4	1	1	1	0	0.80
Mzundu	0	0	3	3	1	1	5	6	0	—	8	1	0	1	2	2.38
Kwamgwe	1	0	0	1	0	1	6	4	0	1	2	1	0	0	0	2.13
Mandera	0	0	0	2	2	0	3	6	0	—	4	0	0	1	0	1.80
Mlembule	0	7	0	0	0	1	3	2	0	—	0	0	0	0	0	1.49
Mkomazi	0	0	0	0	1	2	3	3	0	—	3	3	2	1	1	3.59
Vugiri	0	0	0	0	0	0	2	3	1	—	6	2	1	0	0	1.71
Magoma	0	5	0	2	3	0	4	5	1	3	0	0	1	0	0	2.21
Kisiwani	0	1	1	0	0	1	5	4	0	—	5	0	0	6	0	2.23
Kiwanda	0	0	0	2	0	0	0	4	0	—	0	0	0	0	0	0.50
Daluni	0	1	0	1	0	1	6	2	0	—	3	4	0	0	1	0.70
Mwakijembe	0	0	0	0	0	0	1	1	0	—	1	0	0	1	1	0.46
Maranzara	0	0	0	0	0	1	6	3	0	1	1	3	0	1	1	1.56
Moa	0	0	0	1	0	2	5	5	0	—	2	2	2	2	2	2.66
Total	1	14	6	15	8	12	66	61	2	11[a]	47	19	6	15	8	291
No. of interviews	163	176	176	179	179	178	178	178	178		167	167	167	164	167	
Percentage	0.1	8.0	3.4	8.4	4.5	6.7	37.1	34.3	1.1		34.1	11.4	3.6	9.1	4.8	1.69

Note: No data available for Kwadundwa.

[a]The question about bicycles was omitted from the printed questionnaire (Kiswahili); data on bicycles are thus incomplete.

in 1993 were measurably better off than those we interviewed in 1972. The most likely purchase by someone with money to spare, after meeting needs for food, shelter, clothing, and the education of children, is a radio, watch, or bicycle (Glaeser 1984: 54).[1] This was true in both 1972 and 1993 (although, through an error in typing our 1993 questionnaire—which was mimeographed in Dar es Salaam while we were already in the field—the question about bicycle ownership was omitted). In 1993, a larger proportion (36 percent) of people interviewed owned a radio or watch, compared with about 22 percent in 1972. Another place where expenditures are commonly made is in home improvement. In 1972, 16 percent had a *bati* roof and about 10 percent had a cement floor in their house. By 1993, the percentage having a *bati* roof had risen to 34, while the proportion with cement floors was virtually unchanged. The greatest change in ownership of *bati* roofs occurred in Mzundu. In 1993, two people owned a motorbike, but no one owned a truck; in 1972, two people owned trucks.

Although I do not want to press the comparisons of material culture too far, I will risk making a few comparisons about material possessions between 1972 and 1993. Clearly, installing a corrugated-iron roof on a house costs a lot more than buying a radio. Nonetheless, in 1972, the 238–246 people we interviewed had collectively 299 items, an average of 1.23 per person. In 1993, the 163–179 people we interviewed had collectively 291 items, an average of 1.69 per person. Thus, it can be argued that in two decades, somewhat more things were to be found among the people in the villages we studied. But this still does not represent great wealth—among the 11 people we interviewed in Mwakijembe were owned one watch, one radio, and one *bati* roof (perhaps not surprisingly, the same person owned all three). There was great disparity between villages with respect to average ownership of items (ranging, in 1993, between 3.59 in Mkomazi and 0.46 in Mwakijembe).

There is more to wealth than the ownership of things. Wealth cannot be defined by summing the items we asked about. People have many other demands on their funds, as well as other priorities—school materials for their children, medical treatment, clothing, and so forth. There is only a weak agreement between the values shown in table 2.6 and student assessment of wealth in the villages in 1972 (see chapter 5). There are two anomalies: Kwamgwe had 2.77 items/person in 1972 but was rated 12.9 in wealth; Mkomazi had 2.93 but was rated only 9.8. Even after removing these two villages from the data set, there is only a weak (negative)

TABLE 2.8. Possessions of Those Interviewed in
Both 1972 and 1993 ($n = 98$)

Item	1972	1993
Personal		
Watch	21	31
Radio	21	33
Bicycle	22	6[a]
Tilley lamp	4	5
Cart	0	5
Wheelbarrow	no data	2
Access to tractor	7	6
Maize mill	0	1
Motorbike (*pikipiki*)	0	1
House		
Bati roof	16	28
Cement floor	11	12
Brick house	0	3
Water catchment	5	7
Casement windows	3	4
Total	110	144
No items cited above	60	40

[a]The question about bicycles was omitted from the printed
questionnaire (Kiswahili); data on bicycles are thus
incomplete.

association between student–perceived wealth rank and material possessions. Vugiri and Kisiwani rank high in wealth and in material goods; Minazini, Mwakijembe, Kwamsisi, and Mzundu are at the other end of both
scales (the r^2 among the 16 villages is only 0.144).

As to the question about changes in the material well-being of those we
interviewed in both 1972 and 1993, we were able to make comparisons for
98 respondents. In 1972, 60 people had none of the items we asked about;
in 1993, only 40 people had none of them. For 29 respondents, 1993 was like
1972, in that none of them had any of the items we asked about in either
year. Slightly more items were cited as being owned in 1993 (1.47/person)
than in 1972 (1.12/person) (table 2.8).

Between 1972 and 1993, among those still living, some were "better off,"
some were "worse off," and the situation for some was unchanged. Overall
49 people were better off in the sense that they had more material things
(an admittedly crude measure of well-being). For 31 people, their situation
was unchanged, and for 23 people, their circumstances were worse. Notable
improvement in individual well-being occurred in Mzundu and Moa. On
balance, the people whom we reinterviewed were slightly better off in

TABLE 2.9. Other Sources of Income, 1993

Village	Carpentry	Charcoal	Make dawa[a]	Pots or baskets	Brew beer	Cut or sell wood	Operate duka[b]	Tearoom (hoteli)
Mgera	1	0	1	0	0	1	1	0
Kwediamba	1	2	2	1	0	1	0	0
Minazini	0	1	0	0	1	1	1	1
Kwamsisi	2	0	2	7	3	2	0	1
Mzundu	0	4	2	3	1	5	0	1
Kwamgwe	0	0	0	0	0	2	1	1
Mandera	0	0	0	2	0	0	1	0
Mlembule	1	1	0	2	1	3	1	0
Mkomazi	1	1	1	2	0	3	4	4
Vugiri	1	0	0	5	6	0	1	2
Magoma	1	0	0	3	2	1	0	0
Kisiwani	0	0	1	1	3	0	0	0
Kiwanda	0	0	0	0	7	0	0	1
Daluni	2	0	2	3	0	2	3	2
Mwakijembe	0	1	0	0	0	0	0	1
Maranzara	0	3	2	2	0	6	1	1
Moa	0	0	2	1	0	5	1	1
Total	10	13	15	32	24	32	15	16
No. of interviews	169	171	171	171	168	171	169	169
Percentage	5.9	7.6	8.8	18.7	14.3	18.7	8.9	9.5

Note: No data available for Kwadundwa.
[a] *Dawa* = medicine.
[b] *Duka* = shop.

1993; but the incremental increase in "well-being" in 21 years (34 items spread among 98 people) represents a very small improvement.

In the face of risk, one diversifies. In Tanga, there is more to livelihood than farming (and, where possible, keeping livestock). We counted 171 instances of income-generating activities that help sustain the families we interviewed in 1993. Some families engage in several enterprises, but overall, 100 of the 172 people we interviewed (i.e., those who responded to the questions about income) had some added source of income. Table 2.9 gives the particulars. The largest number engaged in cutting and selling firewood and in making pots and/or baskets; but quite a number of people brewed beer for sale (especially in Vugiri and Kiwanda) or had a shop or tearoom. Several other activities, not listed in the table, were mentioned: making beehives (Minazini), owning and running a maize mill (Mgera), making and selling *maandazi* (a delicious, deep-fried bread; Moa), and making and selling roofing materials (Daluni). Remittances may also be a significant source of income. One person in Mwakijembe noted that her son, who lived in Dar es Salaam and worked for the government, sent money; another received income from a child who is in the timber business (Kwamgwe). There may have been others.

Summary and Analysis

The purpose of this chapter has been to introduce the reader to the main features of livelihood in Tanga Region and to present data that characterize the poverty and levels of material well-being of the people in the villages we studied. There was something dispiriting about returning to Tanga Region in 1993 and revisiting the many villages where we had done our study in 1972. Wherever we looked, we saw much poverty, and in many ways things seemed unchanged and "unbettered" by the intervening two decades. After 21 years it took the same time for women to fetch water as it had in 1972, but obtaining fuel wood consumed large and increasing amounts of time for women. Livestock continued to play a minor role in the local economy. There was apparently no change in the incidence of illness and lost workdays. Overall, the people of Tanga Region may have been a little better off in 1993, but the amount of change discernible over a 21-year period was minimal—disappointing to me and surely to the farmers we interviewed. One feature, to which we turn in chapter 5, is that there appears to be increasing economic differentiation, both among villages and within villages. Thus, some parts of Tanga Region may be developing economically (Daluni, Magoma), while other areas are not (Minazini, Kwediamba). Within villages, economic differentiation among individual farmers has always been present, but it may also have increased between 1972 and 1993.

Biophysical Environment and Local Knowledge

Topography and Geology

Tanga Region consists mainly of rolling lowlands and low plateaus, but with some spectacular, exceptional areas. Most notable are the Usambara Mountains, fault block mountains of Usagaran (Archean) age made of metamorphic gneisses and granulites, which dominate the north-western part of the region. They are divided into two mountain masses: the larger, western Usambaras, which rise abruptly from the surrounding plains to heights of 2,280 m, and a smaller, lower zone, the eastern Usambaras, which attain heights of 1,250 m. Separating them is the broad Lwengera Valley, which stretches northward from the town of Korogwe. There are other mountainous outliers: the Mlinga (1,070 m), Mhinduro (1,035 m), and Mtai (1,060 m) are a series of ranges that lie north of Muheza and parallel the eastern Usambaras. Opposite the massive western flank of the western Usambaras, across the Mkomazi River near Mombo, is a single prominent peak, Mafi (1,480 m). The general level of the terrain rises from the Jurassic age coastal plain sediments on the east (featuring marine terraces formed of coral) to elevations of 500 m near Mzundu, 750 m at Handeni, and 1,000 m at Mgera, a town nestled at the base of picturesque Mgera Hill (1,310 m). The lands west of the coastal Jurassic beds consist of Karroo sediments (sandstones and shales) in an arc of land (see fig. 3.1) stretching from the lower Pangani River, just below Pangani Falls, near Hale, to Mlingano, then northward to the Kenyan border. West of this

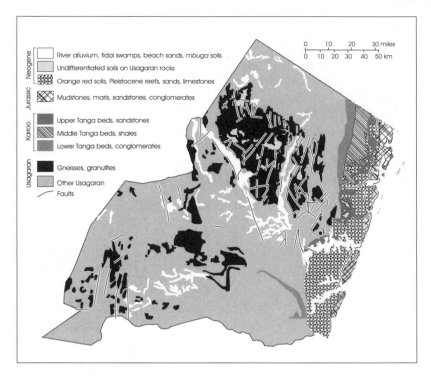

FIG. 3.1. Generalized geology of Tanga Region. (Source: German Agency for Technical
Cooperation 1976a)

fall line, the terrain, made of Archean gneisses and other metamorphosed
rocks, becomes undulating uplands and hills, with occasional granulite or
granitic hills, such as Kwediboma (1,296 m), rising steeply from the local
terrain (fig. 3.2). In the extreme southwestern corner of Handeni District
one finds the foothills of the Nguru Mountains (in Morogoro Region), with
some local peaks rising to elevations of 1,500 m (e.g., Masika, 1,565 m). It
is useful to have a general picture of the geologic layout of Tanga Region
because the terrain affects in major ways the distribution of rainfall and
the sorts of soils and vegetation that have developed from place to place.

Climate

Precipitation

The bulk of our study concerns the agrometeorology of villages at elevations below 500 m. These are villages where rainfall is moderate to

FIG. 3.2. Kwediboma Peak, in western Handeni District. Many isolated granite or granulite hills rise steeply from the general level of the terrain in Handeni and Korogwe Districts. Photo taken 14 February 1993.

low and highly variable from season to season (fig. 3.3).[1] The area has experienced frequent crop failures and food shortages over the years. To live there, farmers must cope with serious management problems caused by rainfall. Two villages in western Handeni District have elevations over 750 m. Only one village is located in montane forest environs, Vugiri, whose nearby rainfall station, Ambangulu, has an elevation of 1,219 m. Table 3.1 gives village names, elevations, and average rainfall values.

Except over mountain masses, rainfall in Tanga Region generally ranges between 800 and 1,200 mm/year (fig. 3.4). Along the coast the amount is 1,200 mm/year, and it declines gradually toward central Handeni to under 800 mm. The Umba Plains in the north, where Mwakijembe is located, have under 600 mm/year. The lowest values in the region are at Mkomazi, less than 400 mm/year, in the northwest, at the gateway to Pare District. In western Handeni District, in the hilly country that includes Mgera and Kwadundwa, rainfall reaches 1,100 mm/year. The wettest parts of Tanga Region are the mountainous areas. Some parts of the western and eastern Usambaras have annual rainfall exceeding 2,000 mm. In chapter 5, rainfall diagrams showing both the monthly average as well as the variability of the

FIG. 3.3. Rainstorm near Mlingano, Muheza District, over a currently inactive sisal estate. Storms are often highly localized. Photo taken early January 1993.

rainfall are provided for the villages. The most salient feature of rainfall is that its seasonal distribution in the eastern part of the region is weakly bimodal, with one strong peak over April or May (the *mwaka,* or *masika,* rains) and the other, somewhat weaker, over November (the *vuli* rains), with a very dry period in January and February. In the far western part of the region, precipitation is essentially unimodal in distribution. Figure 1.1 shows these two contrasts, comparing Moa, on the coast, with Mgera, in the west. At Mgera, the rains peak strongly in April, and the period June through September is dry.

Evaporation

The inverse of moisture supply for crops is moisture demand. Evaporation, suitably weighted by the size and geometry of growing crops, is the key measure of moisture demand in crops. Evaporation values have been based on US Weather Bureau Class A open pan evaporation values for three sites (Tanga, Mlingano, and Mombo). Evaporation rates for other villages have been extrapolated from the nearest of the three sites using elevation as a means of raising or lowering estimated values. These

TABLE 3.1. Village Names, Elevations, and Average Rainfall Values

Village name	Station number	Station name if different	Av. precip. (mm)	Elevation (m)
Kwadundwa	9537004		1,075	853
Mgera	9537001		821	1,006
Kwediamba	9538007	Handeni	837	677
Minazini	9538007	Handeni	837	677
Kwamsisi	9538031		1,103	152
Mzundu	9538038		845	488
Kwamgwe	9538032		1,189	305
Mandera	9538017		906	427
Mlembule	9438022	Mombo	810	415
Mkomazi	9438009	Buiko Railway	382	456
Vugiri	9538004	Ambangulu	2,046	1,219
Magoma	9438016		688	381
Kisiwani	9538025	Kiwanda	1,212	460
Kiwanda	9538025		1,212	220
Daluni	9438027		1,102	366
Mwakijembe	9439026		485	137
Maranzara	9538021	Pongwe	1,163	101
Moa	9439006	Mtotohovu	993	46
Other				
Amani	9538003		1,901	911
Korogwe	9538008		1,047	292
Lushoto	9438003		1,089	1,396
Mlingano	9538011		1,171	188
Muheza	9538042		1,250	198
Pangani	9538006		1,229	9
Tanga	9539000		1,232	9

values have also been compared with estimates based on the method of Penman for other stations in Tanga Region: Amani, Ngomeni, and Tanga (Woodhead 1968). They are characteristically consistent, in that Penman estimates are higher than actual pan evaporation measurements during cooler periods (Porter 1981: 228–229). Figure 3.5 gives an example of the evaporation for Mlingano, using a smoothing function (a 21-day moving average) based on three years' worth of daily observations. In Tanga Region, evaporation conditions are much more conservative, year after year, than are rainfall amounts. Although variations in pan evaporation from day to day can be considerable—as between a windy, bright, sunny day and a still, overcast day—daily average values can be used, because the soil acts as a buffer, providing moisture to crops on high-demand days and little at times of low demand. Average daily values of evaporation range between 7 mm in January and 4 mm in July and August.

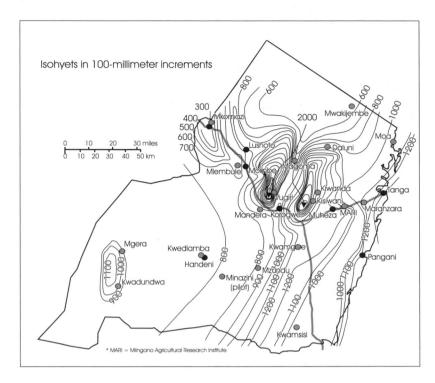

FIG. 3.4. Average annual precipitation in Tanga Region. (Source: Data from Tanzanian Meteorological Department)

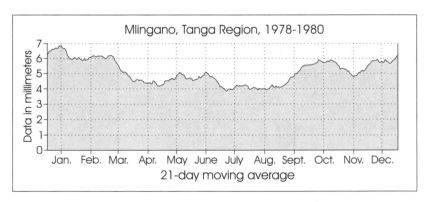

FIG. 3.5. Average daily pan evaporation. (Source: Data from Mlingano Agricultural Research Institute)

TABLE 3.2. The Quality of the Daily Rainfall Record

Site name	Station number	Record length (years)	Total substitutions		Substitutions affecting analysis	
			No.	Percent	No.	Percent
Kwadundwa	9537004	20	46	19.2	4	1.7
Mgera	9537001	35	87	3.1	10	2.4
Kwediamba	9538007	66	10	1.3	5	0.6
Minazini	ditto					
Kwamsisi	9538031	26	51	16.3	6	1.9
Mzundu	9538038	34	18	4.4	4	1.0
Kwamgwe	9538032	33	28	7.0	6	1.5
Mandera	9538017	19	9	3.9	2	0.9
Mlembule	9438022	33	23	5.8	6	1.5
Mkomazi	9438009	27	67	20.7	1	0.3
Vugiri	9538004	63	2	0.3	2	0.3
Magoma	9438016	22	53	20.1	7	2.7
Maranzara	9538021	46	29	5.3	8	1.4
Kisiwani	9538025	39	60	12.8	13	2.8
Kiwanda	ditto					
Daluni	9438027	24	55	19.1	6	2.1
Moa[a]	9439006	66	47	5.9	12	1.5
Mwakijembe	9439026	10	16	13.3	3	2.5
Subtotal					95	1.4
Other						
Amani	9538003	66	12	1.5	0	0.0
Korogwe	9538008	52	41	6.5	3	0.5
Lushoto	9438003	67	1	0.1	1	0.1
Mlingano	9538011	58	0	0.0	0	0.0
Muheza	9538042	36	40	9.3	1	0.2
Pangani	9538006	67	47	5.8	3	0.4
Tanga	9539000	66	23	2.9	4	0.5
Total, all stations					107	0.9

[a] Data for Moa based on nearby Mtotohovu, but for some years data from Mazola (9439063) have been used.

Data Quality

There are 16 rainfall stations under analysis (the stations associated with the 18 villages), although the network of stations in Tanga Region numbers some 51, and other stations such as Mlingano, at the Mlingano Agricultural Research Institute, have been used in analysis (table 3.2). Some of the stations have short or incomplete rainfall records. The longest record for which we have daily values is 66 years (Handeni Town, near Kwediamba and Minazini); the shortest record length is at Mwakijembe, 10 years. (As a general rule, hydrologists and climatologists like to have

11 or more years of observation before they feel able to discuss proba-
bilities.) The average length of record, characteristic of most stations, is
35 years, lying within the period 1955–1992. There are important gaps in
the daily rainfall record. In some cases entire years are missing. In order
to run the model, it is necessary to have a continuous run of data (so
that soil moisture budgets can be continued through dry seasons, and the
like). In any instance where four or more months have had to be inserted,
the simulation results for that season have been ignored. If one, two, or
three months was substituted, the season remains in the analysis. There
were 95 months for which rainfall was substituted, using daily figures from
the same station and months but from other years. Since there were 6,756
station-months in the record of the 16 stations, the rate of substitution is
1.4 percent.

Soils

In each village we interviewed a knowledgeable person at length.[2] We
asked about soil types and their names, characteristics, typical topographic
locations, and uses. Also, 50 sets of soil samples (collected at depth incre-
ments of 15 cm, up to 60 cm) were taken and local vernacular names
recorded. These samples were subjected to laboratory analysis to determine such things as color, texture, pH, available water capacity, and levels
of organic carbon, nitrogen, and phosphorus. From these data emerged 72
soil names, with considerable variation in spellings. Table 3.3 lists the soil
names we encountered, village by village. It is possible to create some order
out of the profusion of terms on the basis of the soil analysis we did, as well
as the way the soils were described by informants as to color, drainage, and
typical site. Despite some ambiguities in the data, including use of the same
term to describe different soils (e.g., Vertisols and black Oxisols/Alfisols at
the base of a slope), we have collapsed the soils of Tanga Region into five
general types (fig. 3.6), four of them along a slope as a catena: (1) Oxisols/
Alfisols on uplands (ferralitic Arenosols), (2) Oxisols/Alfisols on hillslopes
(orthic and rhodic Ferralsols), (3) Oxisols/Alfisols at the base of slopes
(orthic Luvisols and Fluvisols), (4) Vertisols in low areas, and (5) Entisols
on flat land that have been used a long time or exhibit the undeveloped
characteristics of Entisols (from recent).[3] Entisols are of various origins.
For example, some, closer to the coast, developed on sands overlying
coral terraces. Generalizing from local terminology, we could call any soil

TABLE 3.3. List and Classification of Soils Encountered in Tanga Region (Villages Are Arranged West to East)

Village name	Dominant linguistic group	Soil name	Characteristics and locale
1. Oxisols/Alfisols (upland): red			
Kwadundwa	Nguu	*msanga wekundu*	
Kwamsisi	Zigua	*udongo mwekundu*	
Mzundu	Zigua	*mwekundu*	Loam clay
Kwamgwe	Zigua	*udongo mwekundu*	
Mandera	Zigua	*msanga winkundu*	
Maranzara	Digo	*udongo wekundu*	Red soils
Mlembule	Zigua	*udongo mwekundu*	
Vugiri	Sambaa	*shanga ya unguika*	Red soils on hills
Magoma	Sambaa	*waunguika*	Red soils, sticky when wet, on hills
Kisiwani	Sambaa and Bondei	*shanga ye unguika*	Red/brown on hills
Kiwanda	Sambaa and Bondei	*udongo mwekundu*	
Kiwanda	Sambaa and Bondei	*shanga ye unguika*	Red on hills
Daluni	Bondei	*mchanga mwekundu*	Red soils on high and flat land
Mwakijembe	Akamba	*mchanga mwekundu* or *mthangazi*	Reddish, sticky, uplands
Moa	Digo	*chele*	Red soil on hills
2. Oxisols/Alfisols (slope): brown (sometimes black)			
Mzundu	Zigua	*mwekundui*	
Mlembule	Zigua	*msanga ukulugulika*	[red] hills and slopes
Vugiri	Sambaa	*shanga ya kawaidaa*	Brown loam on slopes
Vugiri	Sambaa	*shanga ya chuta*	Black loam in forests
Kisiwani	Sambaa and Bondei	*shanga ya kawaida*	
Kisiwani	Sambaa and Bondei	*shanga ye chuta*	Black on hills
Mwakijembe	Akamba	*mwithanga mwituni*	
Moa	Digo	*shanga ya kawaida*	Mixed soil everywhere
3. Oxisols/Alfisols (lowland): black			
Mzundu	Zigua	*mweusi*	Loam clay
Kwamsisi	Zigua	*kidongo*	
Kwamgwe	Zigua	*udongo mweusi*	
Mandera	Zigua	*kilongo*	
Mlembule	Zigua	*msanga wa kilongo*	In valley bottoms (*mbugani*)
Mlembule	Zigua	*udongo mweusi*	
Mkomazi	Sambaa	*udongo mweusi*	
Maranzara	Digo	*kindongo*	Loam, black, without sand
Vugiri	Sambaa	*shanga ya kiongo*	Clay, warmer places
Magoma	Sambaa	*kilongo*	Black soils on hills
Kisiwani	Sambaa and Bondei	*shanga ye chuta*	
Kiwanda	Sambaa and Bondei	*shanga ye chula*	Black in valleys
Daluni	Bondei	*mchanga mweusi* or *udongo mweusi*	
Mwakijembe	Akamba	*kidongo*	
Mwakijembe	Akamba	*mwihanga mwiyu*	
Mwakijembe	Akamba	*shangashanga*	Whitish, highly infertile basin sides and near *mabonde*
Moa	Digo	*kidongo*	

(*Continued*)

TABLE 3.3. (*Continued*)

Village name	Dominant linguistic group	Soil name	Characteristics and locale
4. Vertisols: black			
Kwadundwa	Zigua	*msanga mtitu*	
Mgera	Zigua	*msanga mtitu*	
Kwamsisi	Zigua	*udongo mweusi*	Black, very slippery during rainy season, big cracks when dry, low water-holding capacity; fairly fertile; bananas, coconuts, maize, and sorghum very suitable for these soils; mostly found in valley bottoms
Mzundu	Zigua	*kilongo*	*mabonde*
Kwamgwe	Zigua	*kilongo*	
Mandera	Zigua	*udongo mweusi*	Blackish with big cracks during the dry seasons; very fertile; suitable for almost all crops grown in this area; mostly in valley bottoms but found in a wide variety of places
Mlembule	Zigua	*msanga mtitu*	In valley bottoms (*mbugani*)
Mkomazi	Sambaa	*shanga ya kilongo*	Black/clay, in valleys
Magoma	Sambaa	*wachuta* or *ktaeuna* (sp?)	Black soils in valleys
Maranzara	Digo	*gandika*	
Kisiwani	Sambaa and Bondei	*shanga ya kinongo*	Clay soils, sticky
Kiwanda	Sambaa and Bondei	*shanga ye kinongo*	Clay, black and sticky, on plains
Moa	Digo	*gandika*	Black soil, small pockets of clay soil
Daluni	Bondei	*mchanga mweusi*	Black soils, valleys and hillsides
Daluni	Bondei	*mchanga wa kirongo*	Clay, in valleys
Mwakijembe	Akamba	*mchanga mweusi*	Black soil, sticky, in drained-to areas
5. Entisols: usually white			
Mkomazi	Sambaa	*shanga ya magadi*	salty/white, in valleys
Kisiwani	Sambaa and Bondei	*shanga ya n'gombe*	white soils, can be used for whitewash, often found under topsoil
Daluni	Bondei	*udongo mweupe*	white
Mwakijembe	Akamba	*mchanga mweupe*	whitish, highly infertile along rivers
Maranzara	Digo	*misanga mwivu*	sandy and black, sandy loam
Maranzara	Digo	*shanga*	whitish and sandy
Maranzara	Digo	*mchange*	sand
Moa	Digo	*fukwe*	light soil mainly near the coast

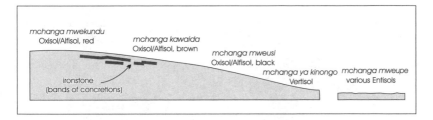

FIG. 3.6. A soil catena for lowland Tanga Region.

udongo (earth) or *mchanga* ("sand" and, by extension, "soil") and add a descriptor: (1) *mwekundu* (red), (2) *kawaida* ("regular" or "ordinary"), (3) *mweusi* (black), (4) *ya kinongo* (after an herb, *Aerva lanata,* that grows on black cotton soils), and (5) *mweupe* (white). These are all standard words in Kiswahili.

Vegetation

Aside from a few forest reserves, there is little land in Tanga Region where the vegetation has not been greatly altered by human activity (fig. 3.7). Many of the forests have been extensively logged in recent years (Kikula 1989: 79). Figure 3.8 shows general vegetative cover types. Montane forest, covering 8.2 percent of the region, is found in four main areas: the eastern Usambaras (fig. 3.9), in two parts of the western Usambaras, a zone stretching from Vugiri to Lushoto with an outlier to the northwest around Mlalo, and, finally, in the isolated southwestern part of Handeni District, centering on the Nguu Kilindi and North Nguru Forest Reserves. These forests are rich, with some 217 tree species identified in the eastern Usambaras (Hamilton 1989: 229). The presence of unique species in isolated patches (e.g., in the high elevations of Mount Mtai) suggests that the forests have been isolated from one another and have existed for thousands of years (Hamilton 1989: 227). The Nguu Mountains, in western Handeni District, are one of the world's 25 "hot spots" for conservation of biodiversity (German Development Service 2003: 3).[4] Species in the high mist forests differ from those of lowland montane areas. According to the Ministry of Lands, Natural Resources, and Tourism (Forest Division 1984), common species in montane forest are *Sonneratia alba* (*mpia*),[5] *Rhizophora mucronata* (*mkoko*), *Ceriops tagal* (*mkandaa*), *Pygeum africanum* (*mueri*), *Serindia usambarensis, Podocarpus* spp.,

FIG. 3.7. Landscape near Vugiri, eastern slopes of the Usambara Mountains, Korogwe District. This landscape was once in montane forest. Over many generations it has been cleared and converted into cultivated fields. Note field-preparation fire in distance. Photo taken 24 February 1993.

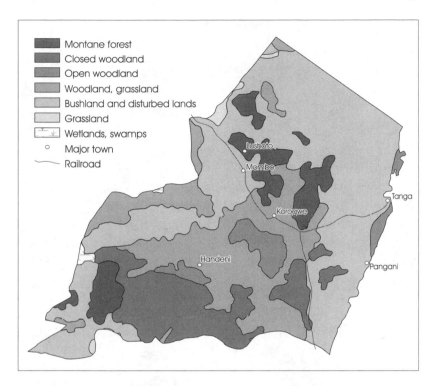

FIG. 3.8. Vegetation cover types based on *Landsat* images and field observation. (Source: Forest Division 1984)

Hypericum spp., *Myrsine africana, Leonotis mollissima, Lycopodium clavatum, Adenocarpus* spp., *Albizzia* sp., *Vitex cuneata (mfudu)*, and *Isoberlinia tomentosa.*

Closed-canopy woodland, covering 13.7 percent of the region, mainly in southern Handeni District, typically contains *Pterocarpus angolensis, Kigelia aethiopica (mwegea), Tamarindus indica (msisi, mkwaju), Acacia polyacantha (mkengewa), Ficus* spp., *Julbernadia* spp., *Brachystegia spiciformis (mrihi), Combretum collinum,* and *Perioria critofolia.* Open woodland, covering 4.2 percent of the region, contains some of the tree species of closed woodland and is characterized by *Tamarindus indica (msisi, mkwaju), Fagara olitoria (mtata), Pterocarpus angolensis, Ficus* spp., *Brachystegia spiciformis, Kigelia aethiopica,* and *Acacia nubica.*

Woodland/grassland, covering 25.8 percent of Tanga region, most of it in Handeni District, is characterized by *Acacia polyacantha (mkengewa), Balanites aegyptiaca (mulului,* Kamb.*), Acacia tortilis (mugunga),*

FIG. 3.9. Montane forest, eastern Usambara Mountains, Muheza District. The all-weather road that passes through Kisiwani and connects Muheza with Amani, was formerly the roadbed of the Tengeri-Sigi rail line. The montane forest is notably rich in the number of species found. Photo taken 8 November 1992.

Commiphora pilosa (*mbambara*), *Solanum incanum, Commiphora* spp., *Acacia* spp., *Combretum gueinzii* (*kiama,* Kamb.), *Baphia masainsis,* and *Maerua angolensis* (*mlala-mbuzi*).

The vegetative type covering the largest part of Tanga Region (47.0 percent) is bushland and disturbed lands. Characteristic tree species are *Acacia mellifera* (*kikwata*), *Acacia nilotica* (*mtetewi*), *Commiphora* spp., *Adansonia digitata* (*mbuyu,* baobab), cactus-like *Euphorbia, Croton* spp., *Grewia* spp., *Combretum* spp., and *Acacia gansfolia.* A small amount of grassland is found on the margins of the region, especially in the lee of the western Usambaras, near Mkomazi, where the associated tree species are *Acacia* spp., *Commiphora* spp., and *Combretum* spp. Locally there are swamps and wetlands covered with a variety of sedges, grasses, and reeds. Many of these wetlands are long, sinuous areas adjacent to rivers and streams and are too small to show on the map. Notable areas of swamp occur along the Pangani, Mkomazi, and Lwengera Rivers.

The large areas of disturbed land became that way through cultivation by local people, livestock grazing and browsing (and associated annual fires), cutting for timber, firewood, and charcoal making, or the extensive alienations for commercial crops, the most notable being sisal estates, which occupy extensive areas in the lowlands (Mascarenhas 1971b: 52). These estates were begun in 1892, during German colonial times, and some fields have been cultivated for more than 100 years, resulting in soils that are greatly impoverished (Hartemink 1995).[6]

Some Examples of Local Knowledge

Farmers not only have considerable practical knowledge of soils and plants but constantly read the environment about them for clues and cues on when to act during the agricultural year. They take cues from plants, from animal behavior, from insects, and from the sky (both the stars and the weather). For example, the *mtondoo* (*Calophyllum inophyllum*), a laurel whose fruit gives an edible oil, is studied throughout the year.[7] The leaves drop in July and August (before the onset of the *vuli* rains—time to prepare the fields), and new green leaves appear in October, indicating the onset of the short rains. New leaves of a reddish color emerge in January prior to the onset of the *mwaka* (*masika*) rains—a time to burn off bush and fields.

The *mbega* (colobus monkey) predicts the advent of the rains once or twice a year by a characteristic high-pitched cry. A bird called *mbizi* (a swallow?) also presages the coming of rain. Finally, the clouds and the state of the atmosphere are watched for signs of coming rain. One cloud type, called *meusingu,* forms in Tanga, east of the Usambaras. These are cumulus clouds that begin to build in the late afternoon and are said to show that rain will come soon.

In all, 157 interviews (from 1972) contained some information about "reading the environment" from the behavior of birds, trees, the weather, and so forth. Over 260 separate items were mentioned. The bulk of them dealt with vegetation (133), followed by bird behavior (53), the state of the weather (40), and insects (12 citations, including ants, crickets, and even insects that purportedly try to enter into one's ears). Among the some 25 birds that were mentioned, we can name several whose appearance (plumage) and behavior (flying patterns and directions, and their songs) are commonly studied: *dudumizi (Centropus,* the white-browed coucal, a cuckoo), *mzingi (Quelea quelea aethiopica,* weaverbird), *kunguru* or *kwambo (Corvus,* crows), and *mbayumbayu* (swallow or swift). Two informants said that a peculiar roar of lions foretold rain. With respect to the weather, clouds and impending thunderstorms are used to predict rain; but how cold it is in the morning, how hot it is at night, the light in the sky at sunset, and wind directions are also used. Further, the positions of stars, especially just at sunset, are studied (14 citations). The people of Moa, a fishing village, use the state of the ocean to give them clues about the weather (5 citations).

Respondents cited over 50 kinds of trees that they studied for clues about the weather.[8] The most commonly studied tree is *mvule (Chlorophora excelsa* or *Milicia excelsa,* East African teak, with 21 citations), but also important are the *mtondoo* (with 8 citations), *msasa (Ficus capreaefolia* and *Cordia ovalismnyina,* 8 citations), and *mkongo (Afzelia guangensis,* 4 citations). Other identifiable trees that people mentioned are the *bambangoma (Erythrina abyssinica), ganga (Euphorbia candelabrum), mwembe (Mangifera indica,* or mango), *mbuyu (Adansonia digitata,* or baobab), *msirisha (Euclea fructuosa,* also called *mdaa* and *misezi), muwale (Raphia kirkii), mfune (Sterculia appendiculata), kihumpu (Mucuna puriens), mkongolo (Combretum schumannii), muungu (Cola usambarensis), mhande (Crabia zimmermannii),* and *mvinga (Pterocarpus bussei).* All of these trees have distinctive flowering behavior, which in most cases becomes evident before the rains have come. Some are diagnostic for the *vuli* rains; others, for the *masika* rains. Some of the clues are subtle. For example, if you plant maize after the flowers of the *msasa* tree have dropped, you have planted it too late. One observes the emergence of sap on one tree, the splitting of pods on another, or watches to see when the *dudumizi,* which has a long beak, goes after snakes.

The foregoing examples show that local people use a wide range of items in their environment to help them predict the weather and to reach decisions on when to do things in the agricultural calendar. In other chapters I will explore still other kinds of local knowledge people have about the environment they use. In chapter 5, details are given for most villages about the crops that farmers grow, the crop calendar (planting dates and season lengths), and characteristic intercropping choices. In chapter 6, I consider how farmers choose which crop to grow, the consequences of delay in planting, their ideas about intercropping or interplanting, and their views and practices on plant spacing. Farmers constantly make livelihood risk decisions, trading off between quantity of production and reliability of production. In chapter 7, I examine the history and causes of food shortages and famines in Tanga Region, together with how people (both institutionally and individually) have tried to cope with them. In each of these chapters I show that there is a complex interplay between Western scientific knowledge and local knowledge.

CHAPTER FOUR

Tanga's Regional History

In this chapter I present salient features of Tanga Region's population and culture history, its environmental history, and its political economy. It is a large order, but to understand the people of the 18 villages we need to know what they are like and what has gone on in their past that impinges on their food security and efforts to fashion livelihoods. With this background in hand, we can appreciate what the people have done in relation to the population growth Tanga Region has experienced over half a century, and we can better understand farmer local knowledge and practice (chapters 5–7). The chapter is presented in two parts. The first is about the historical events that have affected the people of the region at various times from about 1850 to the present. The second is about the demography and culture of Tanga Region and the people of the 18 villages whom we interviewed.

Historical Events

Any analysis of Tanga's agriculture and economy needs to be set in the context of the region's long and complex history. This is a region whose people have engaged in local and long-distance trade for hundreds of years. We find trade along the famed Azania, or Mrima, coast in the 10th century, and perhaps even as far back as AD 100 (Chittick 1974: 104–105; Oliver and Mathew 1963: 93ff.; Datoo 1970). Yet it was during the past 200 years that the distinctive regional character of coastal Tanga and Pangani and their hinterlands formed. Trade in ivory and slaves, as well as in foodstuffs, was well established by 1811 and this gave rise to a relatively prosperous regional economy (Giblin 1992: 22).

Key events in the 19th- and early-20th-century history of Tanga region include (1) the development of the clove industry in Zanzibar in the 1830s and 1840s, its attendant need for labor, and thus the increased pace of the slave trade; (2) the competition after 1840 from well-capitalized Zanzibari merchants, who increased the size and range of caravans and took trade away from local Zigua traders, and who developed plantations along the Mrima coast; (3) the German conquest of 1889; (4) the rinderpest epidemic of 1891 and its consequences; and (5) from 1904 onward, the development of the sisal industry.

Between 1891 and 1956 the most dramatic deterioration of livelihoods was caused by the spread of bush habitat for tsetse fly, resulting in a steep decline in cattle keeping, as well as disruptions in agriculture. The resulting loss of environmental control by Zigua farmers and livestock keepers led to a decline in numbers of people. The vegetation, no longer managed each year by fire and cutting, changed its character. This, in turn, led to increases in tsetse fly infestation and trypanosomiasis in cattle and to the spread of ticks (*Theileria*) and tick-borne diseases such as East Coast Fever (Ford 1971; Porter 1976; Kjekshus 1977; Giblin 1992; Lambrecht 1964). Figures 4.2 and 4.3 show the dramatic increase in tsetse fly infestation between 1913 and 1955. I discuss each of these topics in turn.

Transport, Circulation, and Isolation

We need to take account of one fundamental feature of lowland Tanga. The bulk of movement was and continues to be on foot. Perceptions about a place are easily biased by how one reaches it (figs. 4.1 and 5.15). A Tanzanian colleague observed to me: "You know, our Tanzanian roads are really comfortable, if you just get out and walk on them." Since I have usually moved from one place to another in Tanzania by vehicle, I have a very different idea about good roads and bad roads. I am struck, in reading about Handeni in the 19th century, at the importance of the trade links to the Nguu (or Nguru) Mountains. This is where Kwadundwa is located, the village we could barely reach in 1972 and which we were not permitted to visit in 1993 because the roads were so impassible. Places like Kimbe and Kwekivu, in the heart of the Nguu Mountains, were important caravan stops in the 19th century (Giblin 1992: 23, 51).

FIG. 4.1. Road to Mzundu, Handeni District. This road is impassable for vehicular traffic during the rainy season. Photo taken 22 February 1993.

Yet, if local people and caravan traders circulated easily within the region, Tanga and particularly Handeni District were, from the standpoint of colonial administration, inaccessible. For much of its colonial history, Handeni has been off the beaten track—difficult to get to and difficult to move around within. For example, in the early 1970s, one still reached Handeni from Dar es Salaam via Morogoro, a trip of 378 km, whereas the distance as the crow flies between the two places is 241 km. Tanga Town, the provincial headquarters, which lies 193 km north of Dar es Salaam, could be reached by boat, and after 1963 by train, but the road journey was 563 km, much of it over bad roads. Within the district, movement, particularly during the rainy season, was difficult and sometimes impossible. The administrative headquarters at Kwekivu in the southwestern part of the district was usually reached by traveling west to Mgera and then south along the edge of the Maasai Steppe in the rain shadow of the Nguu Mountains (John Ainley, pers. comm., 14 July 2002). This was a distance of over 130 km, whereas the direct distance from the town of Handeni was 87 km.

Randal Sadleir described efforts to inform aviation authorities in Dar es Salaam about the crash of a Central African Airways Viking plane in 1953 in eastern Handeni District, north of Kwamsisi at a place called Kwedifunda: "There was no radio, telephone or telegraph anywhere near. There was apparently some kind of signal station at the little port of Sadani on the Indian Ocean in the northern Bagamoyo district where it might be possible to 'tap into' the coastal telephone line from Tanga to Dar es Salaam, and, we hoped, get a message through to headquarters" (Sadleir 1999: 137–138). By driving at night, he and David Brokensha succeeded in getting the message through only two days after the crash itself. The official inquiry found the cause of the crash to be metal fatigue—one wing had come off in turbulence. All 13 people aboard were killed.

Still further evidence of isolation is found in the degree to which local administrators were supposed to run their own show. Sadleir summarizes well his many responsibilities as a district commissioner:

> In addition to my main tasks of representing the sovereign, maintaining law, dispensing justice, collecting revenue and ruling the district through chiefs, I was responsible for various other duties there. I was the administrator-general dealing with bankruptcies, the registrar of births, deaths and marriages, the coroner, the police (prisons) probation officer, and a representative for the agriculture, education, forestry, game, grain storage, labour, medical, veterinary

and water development departments. I was the postmaster, executive officer for the Chanika township, chairman of the district team and a representative of the departments of civil aviation, customs, hides and skins, public relations, public works, social development and tsetse fly. I was also responsible for the census, resettlement schemes and for liaising with missions. (1999: 143)

The Clove Industry in Zanzibar and Zanzibaris on the Mainland

Although Arabs had traded along the coast for centuries and maintained permanent settlements since the 10th century, the Portuguese controlled the Indian Ocean coast from Mombasa to Sofala until the end of the 17th century. In 1699 they were ousted from Fort Jesus in Mombasa, and thereafter control was contested between Arab and local coastal groups, such as the Mazrui of Mombasa. Commercial development and trade began to grow when Seyyid Said of Muscat (and Oman) made Zanzibar his capital in 1828 (Lofchie 1965: 32). Said created a strong state that lasted for 60 years and exercised control over the islands and many coastal towns. At one time three countries (Great Britain, France, and Germany) had consular offices in Zanzibar.

Omani Arabs, in developing the clove industry in the 1830s and 1840s, occupied the largely uninhabited, malarial, dense tropical forests of the western half of Zanzibar (Middleton 1961: 11). The indigenous people were forced to live in the poorer eastern areas (Lofchie 1965: 48). Clove production has high labor demands, and this led to an increase in the slave trade on the mainland, since labor for the clove industry was provided by slaves. Estimates of the resident population who were enslaved vary between two-thirds and three-quarters (Lofchie 1965: 490). The Arab plantation owners after a time became heavily indebted to Indian merchants; by the 1880s two-thirds of the plantations were "fully mortgaged to Asians" (Lofchie 1965: 105).

On the Tanganyika coast, local traders after 1840 began to feel the pinch of competition from well-capitalized Zanzibari merchants (Giblin 1992: 45). Zanzibari traders were in search of ivory and additional slaves for the clove plantations.

In addition, Omani Arabs from Zanzibar developed sugar plantations along the Pangani River, upriver from the coastal town. These plantations used slave labor exclusively. The owners of these plantations tried to introduce a particularly brutal form of slavery, "chattel gang slavery," which was quite different from the other forms of slavery that existed in the region

(Glassman 1991: 297). Glassman provides a subtle discussion of the nature of slavery along the Mrima coast and its hinterland.[1] Slavery was for most not an either/or situation; it was a state of being continually contested between slave and master. It can be seen as an extension of other forms of dominance and subservience in local society. "According to widespread consensus, seniors had the right to dominate juniors, males had the right to dominate females, and patrons could expect deference from their clients. The language of slavery was but a set of variations on these themes" (Glassman 1991: 295). Resistance from those enslaved culminated in 1873 in the successful Mauya slave rebellion. One of the leaders of the revolt was Akida Bushiri (Abushiri bini Salim). A number of "maroon," or *watoro,* communities, such as Makorora, were established along the Mrima coast. One notable feature of such revolts is that those who had gained their freedom did not reject the dominant Swahili culture. Rather, they wished to participate in it more fully, as followers of Islam, as cultured Waswahili, and even as slave owners themselves (Glassman 1991: 305–306).

German Conquest

Although the German missionaries Johann Ludwig Krapf and John Rebmann had traveled and documented a great deal about coastal and interior East Africa for a German audience in the 1860s, Germany had no official connection with the area. Indeed, Krapf and Rebmann were not acting for Germany at all but were missionaries for the English-based Church Missionary Society. In 1884, Dr. Karl Peters and two associates traveled to the Usagara Mountains (southwest of Kilosa) and signed treaties with local chiefs on behalf of the German Colonization Society. This was followed by other excursions and treaty signings by Peters and others as far south as the Ruvuma River, inland to Lake Tanganyika, and northward along the coast as far as the Tana and Juba Rivers.

The sultan of Zanzibar protested, but to no avail (and he got no help from the British), and with the arrival of German military and naval forces, Germany established control over a large part of Tanganyika. The Anglo-German Agreement (November 1886) ostensibly gave the sultan of Zanzibar continued control over the coast to a depth of 10 miles, plus the islands of Zanzibar, Pemba, Mafia, and Lamu, but not the interior. By April 1888 agents of the German Colonization Society had persuaded the sultan to lease to them his rights to the coast for 50 years in exchange for certain payments. The German Colonization Society became the German

East African Association the following August and assumed control of the entire coastal area of what is now Tanzania.

The coastal people deeply resented this turn of events and opposed the Germans everywhere. One expression of animosity in Tanga Region was the Bushiri (or Mrima) War of 1889 (Giblin 1992: 87). It takes its name from the same Akida Bushiri who had helped lead the successful 1873 revolt against the Omani Arab sugar plantation owners. For a time, the company lost control, abandoning all of the coast save Bagamoyo and Dar es Salaam (Gwassa 1969: 105–108). Since it was clear that the company could not administer the territory, in October 1889 Germany declared the land to be an Imperial German Protectorate. Captain Hermann Wissmann commanded a military force that brought the countryside under control by mid-1890.

German rule between 1889 and 1907 was very repressive and harsh, featuring corvée, or forced, labor, mandatory cultivation of cotton, and other rules and restrictions. The German colonial administrators were aided by the *akidas,* salaried African officials (Giblin 1992: 114–116; Kurtz 1978: 7). Marcia Wright comments that one developmental aspect of German administration was the decision to promote Kiswahili as the language of administration and to train Africans to be civil servants (1968: 625). Schools were established that were staffed by Muslim teachers. The first such school was in Tanga Town in the 1890s. The German administrators had the somewhat naïve idea that the well-developed commercial networks could be converted into an administrative network. Nonetheless, the choice of Kiswahili made it possible to exert control sooner, and the *akida* schools of Tanga Region produced many African civil servants.

The *akidas* were Kiswahili-speaking African and Arab Muslim men, most of them from the coastal area or Zanzibar. They occupied an administrative position interposed between the colonial government (both German and British) and the local African population. Their role was to ensure that local people knew and followed rules and regulations. They kept records, collected taxes, aided the police, adjudicated minor disputes, kept the roads in repair, and guarded governmental property and forest reserves, ensuring that people did not cut trees or kill protected game. They oversaw the work of *jumbes* (village heads). Tanga District, for example, was divided into 11 *akidats.* The 11 *akidas* supervised 123 *jumbes,* who were in charge of single villages or groups of villages ("Native Administration," *Tanga District Book,* reel 1: 9–19, Tanzanian National Archives). According to James Giblin: "Although the Akidas had power, they possessed little of the legitimacy that precolonial leaders had derived from kinship, marriage alliances, and service as patrons, healers, and ritual leaders"

(1992: 115). *Akidas* were permitted to operate Qur'anic schools and to build mosques. They played an important role in bringing Islam to coastal areas and the interior (Giblin 1992: 115).

African resistance culminated in the Maji Maji Rebellion (1905–1907), a mass movement involving perhaps four million people, although it did not spread to Tanga Region. Its root cause was German conscription of male labor for cotton plantations and other enterprises. This led to struggles within the household. Men's labor was diverted, and as a result women became responsible for local economy. This diminished the power of the local male *jumbes*. The economy entered a downward spiral, with great destruction of crops by bushpigs and consequent famine and death from starvation. In the turmoil created by the rebellion, there was loss of control of labor in local societies all along the coast from Mijikenda in the north (Kenya) to Zaramo in the south, as young men left their villages to seek wage work (Willis and Miers 1997; Sunseri 1997: 250ff.).

The rebellion was suppressed by German forces "with the utmost barbarity. Large districts were denuded of their food supplies and the crops burned; the natives who had escaped from devastated areas were driven back again to their homes, where some thousands died of actual starvation" (British Foreign Office 1920c: 47; see also Sayer 1930: 75; Gwassa 1969: 85ff.). Ultimately, through direct killing and the ecological and economic disruption and starvation that followed, an estimated 120,000 Africans lost their lives (Iliffe 1969: 9ff.).

In the aftermath of the rebellion, German colonial policy was rethought and "renegotiated" with the indigenous population. The new policy ultimately affected Tanga Region. The policy, formulated by Dr. Bernhard Dernberg, secretary of state for the Colonial Department, was introduced in 1907 by Albrecht Freiherr von Rechenberg, governor of German East Africa (Listowel 1965: 47). It was supposed to introduce "enlightened economic imperialism," a policy grounded in peasant production of indigenous crops using locally established methods (Iliffe 1969: 76). This approach was designed to ensure security and economy for the local populace (Iliffe 1969: 3). In the period from 1907 to the outbreak of World War I, the policy was vigorously contested by an increasingly politically adroit settler population, which numbered 2,570 by 1906 (Iliffe 1969: 57).

Rinderpest

According to Giblin, there is amply documented evidence that there was a severe population decline in Handeni District from 1890 to about 1940, one consequence of which was a near total loss of livestock. Many formerly cultivated areas were encroached upon by bush and tsetse fly. "The disease environment of the 1940s was the product of a process which weakened social control of vegetative communities, wildlife, vector, and trypanosomiasis" (Giblin 1990: 79).

The cause of this population decline and loss of control of environment was the rinderpest epidemic of 1891.[2] Rinderpest (German for "cattle plague") is a viral disease spread by air and contact. Before 1884, it was unknown in Africa south of Egypt. It came to tropical Africa in 1889, probably brought by horses or cattle used by the Italian army in its occupation of Eritrea (Ford 1971: 138). The disease spread rapidly. In 1890 and 1891 it swept through East Africa and by 1896 had reached the Cape of Good Hope. Wild animals as well as livestock suffered severely: buffalo, eland, bushbuck, giraffes, as well as warthogs, bushpigs, and forest hogs died in large numbers. It was less severe among hartebeests, waterbuck, zebras, elephants, and hippos. Mortality among livestock was estimated to be 95–99 percent (Kjekshus 1977: 129ff.).

The disruption to people's livelihood at this time was great because not only did rinderpest wipe out the subsistence base of many people, but it was followed by smallpox (Giblin 1992: 127; Kjekshus 1977: 132). Hundreds of thousands of East Africa's people died in the decade of the 1890s. Further, the death of such large percentages of stock and wildlife changed the vegetation. Forest, bushland, and thicket claimed large areas that had been open woodland and grassland. There were far fewer animals grazing and browsing, and the annual firing of grazing areas was interrupted. The environment for tsetse fly improved and the fly spread rapidly. Game animals, which are tolerant of trypanosomiasis, recovered more quickly than the stock. A stock census carried out in 1902 in German East Africa gave 460,000 cattle. If stock mortalities had been 95 percent when the rinderpest epidemic passed through, the national herd some 10 or 11 years earlier may have been over 9 million. The national herd in Tanganyika was about 5 million in 1961, 8 million in 1971, and nearly 14 million in 1990 (Central Statistical Bureau 1968: 86; Porter and Flay 1998: 47). "Precolonial societies did not eradicate tsetse populations, but instead coexisted with vector and trypanosomes" (Giblin 1990: 70). The cattle were kept in relatively fly-free areas, but in the 1890s, in the wake of successive famines, "agriculturalists lost their ability to regulate vector-bovine contacts, and cattle virtually disappeared" (Giblin 1990: 70). Human trypanosomiasis, or sleeping sickness, did not occur in Handeni District and lowland Tanga

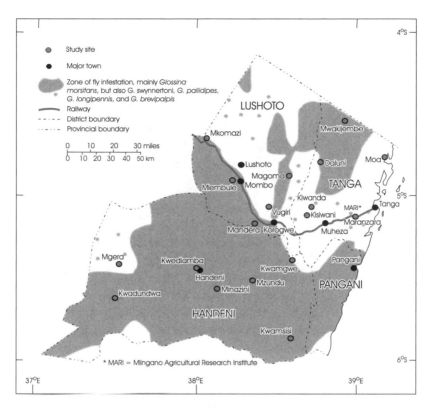

FIG. 4.2. Areas of tsetse fly infestation, Tanga Region, 1955. (Source: *Atlas of Tanganyika* 1956)

Region (Giblin 1992: 30), though there were serious outbreaks elsewhere in Tanganyika in the first decades of the 20th century.[3]

The distribution and purported advance of the tsetse fly and efforts at remediation are traced in figures 4.2 and 4.3.[4] Questions have been raised about the accuracy of some the maps on which figure 4.3 is based (Bax 1943; Giblin 1990: 67). A map of 1909 from the *Deutches Kolonial-Lexikon* (not reproduced here) shows a belt of fly infestation stretching across lowland Tanga Region along the course of the Pangani River. The situation in 1913 shown in *Die deutschen Schutzgebiete in Afrika* has an added fly zone in western Handeni District, over the Nguu Mountains (this is one of the problematic maps). By 1937, as shown in a map by W. H. Potts, fly has covered all but the western part of Handeni District, all of Pangani District, and major portions of lowland Lushoto (northern Umba Plains) and Tanga Districts. Further advances of the fly belt are shown in the tsetse

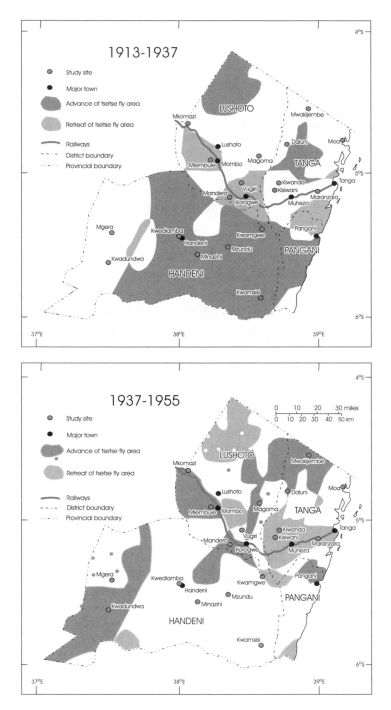

FIG. 4.3. Advances and retreats of tsetse fly zones. (Source: Schnee 1920; Reiches-Kolonialamt 1914; Potts 1937; Survey Division 1956)

fly plate in the *Atlas of Tanganyika* (1956: 12). Advances and remissions (the latter commonly as a result of governmental vegetation clearance programs) are worth showing for two periods, 1913–1937 and 1937–1955 (fig. 4.3). In the first period there is notable clearing of fly infestation in the rapidly developing area of sisal estates, while expansion of tsetse fly occurs in Handeni and Pangani Districts and on the Umba Plains. In the second period there are fly advances in northern Tanga District (around Mwakijembe) and in the mountainous part of western Handeni District. A zone of fly advance is also seen between the towns of Handeni and Korogwe. Overall, fly infestation and risk of trypanosomiasis have limited cattle keeping over much of lowland Tanga Region.

Sisal

Sisal, more than any other crop, has had the profoundest impact on the people and landscapes of lowland Tanga (Mascarenhas 1970). In the early years of the German occupation, colonial officials were interested in developing a crop that would do well in the lowland area, where soils were poor or of indifferent quality, and where rainfall was erratic and highly unreliable. Sisal, a material used for making rope, twine, mats, and rugs, was identified in 1892 by a German agronomist, Dr. Richard Hinsdorf, as a crop that might be suitable.

Sisal (*Agave sisalana perrine*) is found in lowland Mexico, including the Yucatán Peninsula. Hinsdorf discovered that it was illegal to export sisal from Mexico, so he arranged for some seeds to be smuggled out. Only 62 of the 1,000 bulbils sent via Florida and Hamburg survived the journey, but it is from that small stock that the planting material for East Africa's sisal industry was propagated.

Sisal is a labor-intensive and bulky crop and thus its development required good transport facilities and ample labor. The first railway in East Africa was built from Tanga to Moshi between 1891 and 1911. Construction took a long time to get organized, and the engineering of the line was poorly done at first (Hill 1962: 63). The line reached Muheza in 1896 and then stagnated for some years, although the roadbed had been cleared to Korogwe (Hill 1962: 63–64). The railway reached Korogwe in 1902, Mombo in 1904, and Mkomazi (Buiko) in 1910 (British Foreign Office 1920b: 45). Thus, a zone of lowlands paralleling the drainage of the Pangani and Mkomazi Rivers was served by rail after 1910 (Hill 1962: 74; Mkama 1969: 11; Gillman 1942: 19).

Narrow-gauge tracks along which carts loaded with cut leaves could be pulled were even developed on some sisal estates. Alternatively, tractors pulling carts were used to haul the leaves to the decorticator. The decorticator is a huge machine with rotating drums that strip the leaf pulp away from the fibers. The ratio of pulp to fiber is between 20 to 1 and 40 to 1; that is, a metric ton of leaf results in only 25–50 kg of fiber. It takes a great deal of water to run a decorticator, another consideration in locating the sisal-processing facility. According to Mascarenhas, the average sisal estate required 4,000 gallons/hour to wash the fiber and remove the waste leaf pulp (Mascarenhas 1971b: 52).

The labor requirements of sisal are also substantial. In plantations sisal is commonly set out in large blocks in densities ranging from 2,500 to 9,000 plants per hectare. After land preparation and after the bulbils have been transplanted from the nursery, much manual labor has to be performed—weeding, hoeing, desuckering—before any leaf is harvested. Cutting of leaves can begin when the plant has reached two years of age and continue for up to 10 years (fig. 4.4). Eventually the plant "poles," that is, puts up a tall central stalk containing bulbils, which when shaken from the plant take root themselves (fig. 4.5). Sisal can be grown from seed or bulbils, but

FIG. 4.4. Sisal, freshly cut and stacked, ready to be taken to the factory. This photo was taken east of Maranzara on the road between Pongwe and Tanga, in Tanga District, on 18 March 1993.

FIG. 4.5. Old, "poled" sisal near Magoma, on an abandoned sisal block, Korogwe District. The bulbils shake readily from the branches, giving rise to new plants. Photo taken 25 February 1993.

transplanting bulbils is the normal practice. After the fiber has been separated, it has to be set out on racks to dry before being bulked for shipment or used in local manufacture.

Although the developers of sisal estates probably thought they could draw labor from the surrounding areas, events proved them wrong. They found that local people were more interested in growing and selling food to the sisal plantations than in working on them, just as in the earlier caravan trade, they had provided food for traders and in local markets (Feierman 1974). A typical sisal estate might employ 1,500 people. The people lived in company-built and managed housing, "labor lines." There was constant turnover as individuals completed their contracts, usually six months, but sometimes as little as 30 days (the *kipande* had 30 places where marks or punches for each day's work could be recorded).

The sisal industry was in German hands until 1914. After the end of World War I, the sisal lands were auctioned off by the Custodian of Enemy Property. A significant portion of the sisal estate land, about one-third, was bought by Greek entrepreneurs; British and Asian growers took a quarter each; Swiss and Dutch interests bought about six percent each (Mascarenhas 1971b: 52).

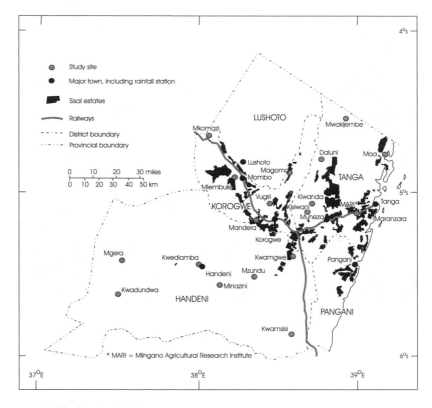

FIG. 4.6. Distribution of sisal estates, c. 1990.

    The sisal industry grew from small beginnings, after 1910 (figs. 4.6 and
4.7). Sales were interrupted during World War I but resumed in 1918 (with
a small peak, as sisal stored during the war was sold off in 1919 and 1920).
A similar situation occurred during World War II (1939–1941), when production declined. Expansion was strong and rapid in the two decades following World War II. On the eve of Tanganyika's independence (1961),
there was great expectation that sisal would be a significant source of
foreign-exchange earning. Indeed, Julius Nyerere thought that along with
tourism and diamonds, sisal would be the mainstay of Tanganyika's economy once independence had been achieved (Engelhardt 1962: 112). The
industry had strong multiplier effects, the most notable of which was that
local people grew food to sell to the sisal estates, which had tremendous
requirements. Near its peak, in 1952, the sisal industry employed 142,000
laborers (Mascarenhas 1971: 52).

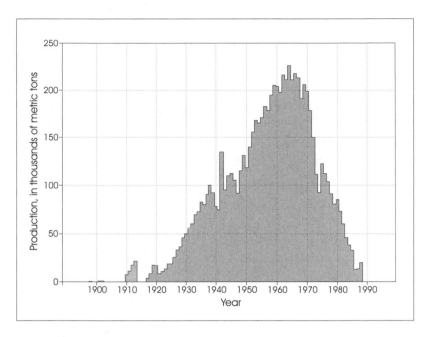

FIG. 4.7. Sisal production. (Sources: Various, see text)

Although sisal plantations developed along both rail lines, that is, along the Tanga–Arusha line and the Dar es Salaam–Kigoma line, the bulk of the development was from Tanga to Korogwe to Mkomazi, with lesser centers near Morogoro and Kilosa (Mascarenhas 1971b).

The sisal industry reached its zenith in production in 1964 and declined thereafter (fig. 4.7).[5] Some 230 million metric tons of fiber were produced in 1964, valued at T Sh 437 million (about $250 million, in constant 1989 US dollars). Sisal is subject to severe fluctuations in price and competition from synthetic fiber. The years since the mid-1960s have been marked by a steep and steady decline in the amount produced, and even more so in the value of the crop. In 1989, production was only 19.6 million metric tons, and its value in 1989 US dollars was $4.27 million (Mpigachapa wa Serikali 1990: 44, 161). The landscape reflects the devastating decline of the sisal industry, with abandoned blocks (fig. 3.3) choked with a growth of brush and relic sisal plants that have "poled," rusted, abandoned machinery, and empty, derelict buildings where laborers once lived. In 1993 there were some signs of revival, and blocks near Hale had been planted out and new housing for workers and staff was under construction. A full revival of the industry will be possible only if world prices for sisal increase and remain high.

British Colonial Policy

The British governed Tanganyika under a mandate from the League of Nations from 1920 and, after World War II, as a trusteeship from the United Nations from 1945 to 1961. Under the mandate, the interests of the African population were to be paramount. The colonial government, thus, had to deal with pressures from a small but well-organized settler community that wanted to grow in areas of high potential. Some of the areas of European settlement had already been well established in German times—Arusha, Meru, Kilimanjaro, Oldeani, and the southern highlands (Iringa, Mufindi, Mbeya, Mbozi, Rungwe, and Njombe). With respect to Tanga Region, the bulk of land alienations were in the lowlands, where sisal was grown; but there had been significant alienations during the German colonial period in the high-potential country of the Usambaras. These alienations were at high elevations, particularly tea estates around Amani and in the western Usambaras around Lushoto and Soni, and west and north of Vugiri. I estimate that 33,000 ha were alienated in the highlands of the eastern and western Usambaras and another 70,000 ha set aside as forest reserves in these areas. These alienations continued to be under European control, but after the 1920s, little new land was alienated. In addition, at least 15 companies, mainly Sikh and other Asian groups, were active in logging forested areas in the Usambaras in the 1950s and 1960s (Central Statistical Bureau 1967).

In general, British policy was minimalist (the "night watchman state"): keeping the peace and ensuring that people did not starve and that society did not change greatly. The Tanga lowlands were seen by the British colonial government as having low potential and as a problem area, prone to droughts and food shortages. Government policy was to try to get people to plant drought-resistant crops, and it promoted such cash crops as cotton and (in high-potential areas) coffee. The need for cash was necessitated by the strict enforcement of poll taxes and hut taxes payable only in shillings (Giblin 1992: 156). This made it necessary for farmers to sell in markets to obtain cash. After 1920, under the mandate, the British decided "against [promotion of] industry, supposedly in an effort not to 'detribalize' the country" (Sarris and Brink 1991: 22).

British policy with respect to food shortages was reactive and palliative; it was not devoted to finding ways to improve the economy, productivity, and the lives of the people. It had no overall plan to improve agriculture beyond a belief that if it encouraged the development of yeoman farmers who used modern methods and marketed their crops, the demonstration effect would trickle down and encourage poor farmers to do likewise.

For much of the colonial period, there was an explicit policy of preventing social and economic differentiation and rural accumulation. The Credit to Natives Restriction Act prohibited institutions (banks) and individuals (traders) from lending to Africans, on the grounds that this would keep them out of debt (Sadleir 1999: 118). The British refused to sell trading licenses to chiefs or to make credit available to them. The British wanted chiefs to be solely salaried servants of the colonial administration (Giblin 1992: 150–151). There were efforts to get farmers to plant crops that would ensure their food security. For example, each farm family in Handeni District was required to plant a plot of cassava (Ainley 2001: 113). There were also programs promoting cotton and cashew tree cultivation, the idea being that the crops would be sold and infuse money into the local economy (Ainley 2001: 152ff.). In promoting cotton without changing the tools used or the labor power of the farmers, British policy created a tension in livelihood. Labor devoted to cultivating cotton (a labor-intensive crop with critical timing with respect to labor inputs) could not be devoted to growing food. One result of British policy was that sorghum production, which is well suited to Handeni's erratic rainfall, declined. In 1993, practically no one we interviewed mentioned sorghum in describing his or her crop calendar.

Sorghum uses about as much water as maize and takes about as long to grow. Its advantage is that it is drought resistant and can survive periods of moisture stress that do irreparable damage to maize. Its disadvantage is that the open panicles of grain are defenseless against birds. Thus, much time has to be spent guarding the crop from birds as it matures. A simulation of 150-day maize and 150-day sorghum grown under the same conditions in Handeni showed that sorghum had an advantage of about 24 percent; its yields were more reliable and usually higher owing to its ability to withstand periods of moisture stress.

With the Native Authorities Act of 1937 British policy toward peasant farmers became more coercive—specifying minimum acreages for certain crops, requiring soil conservation work, assuming greater control of the cooperative movement, and even specifying working hours (Sarris and Brink 1991: 23). Conscript labor was used during World War II. Other governmental programs, such as the Mlalo Rehabilitation Scheme of the 1940s, impinged on the autonomy of farm families. In areas with steep slopes, farmers were required to undertake tie-ridge terracing and other forms of soil conservation and sometimes abandon fields on steep slopes altogether (Scheinman 1986: xiii). Areas near streams were made off-limits for certain kinds of uses (cutting firewood, cultivation, even grazing). The work of terracing was physically arduous and very time-consuming and was most often done by women (Scheinman 1988; von Mitzlaff 1988: 10; Feierman 1993: 138).

Among the Sambaa was a tradition of permitting poor, landless people to grow crops on land not currently being used. As land came to be improved with tie-ridging, it was no longer available to the poorest people for cultivation but became "permanently subject to rental" (Feierman 1993: 114). Now, rich farmers lent land to poor landless farmers, who had to do the work of tie-ridging. The next year the owner took the land back and offered the landless farmer another plot, which he then had to tie-ridge (Feierman 1993: 139). The Usambara Scheme, begun in 1951 and suspended in 1957, "led to a full-scale farmers' revolt." It required every taxpayer to put in "a half acre of tie-ridges." It forbade burning of plant residues, forbade cattle grazing on stubble in harvested fields, and required the planting of elephant or Napier grass (*Pennisetum purpureum*) along all water courses (Feierman 1993: 136–137).

In drier areas there were rules regarding fire, hunting, woodcutting, and stock raising, as well as interregional movement of crops and other items for sale. The British maintained tightly regulated markets with limits on prices and the amounts that could be sold (Giblin 1992: 157). The colonial officers always had a rationale for their policies. For example, the prohibition of movement of foodstuffs supposedly ensured local self-sufficiency and prevented Asian merchants from exploiting local farmers. The policies in aggregate, however, tended to constrain farmers as to what they could do to ensure their food security and improve their well-being.

In due course, these policies helped politicize the rural population and increase awareness of wider social movements in Tanganyika. The hard work of terracing was particularly resented in the Usambaras and elsewhere (Feierman 1993; Young and Fosbrooke 1960). The fact that Tanganyika was a United Nations trusteeship enabled nascent political movements (the Tanganyika African Association and then the Tanganyika African National Union, predecessor to Chama Cha Mapunduzi) to internationalize the struggle for independence (Ogot 1974: 306). Throughout the British colonial years the League of Nations or the United Nations periodically sent commissions of inspection or investigation (Iliffe 1974: 306; Spear 1996). In the latter years of the British period, as independence approached, the colonial administration began to change its policies. The result envisaged is expressed in the following statement by D. W. Malcolm, principal assistant secretary, Agricultural and Natural Resources:

> I want to see the emergence from our hitherto undifferentiated African society of a substantial number of rich men. . . . I would like to see men in a sufficiently strong financial position to be able to send their sons overseas for education, to afford motor-cars, good houses and the like, and I believe that the emergence of such relatively wealthy individuals in the community will provide a stabilising factor of immense importance to the future of this country. (Cited in Iliffe 1971: 37)

Government-sponsored experimental settlement programs were also instituted (Kates, McKay, and Berry 1970; Berry and Kates 1970).

In reading colonial reports and retrospective memoirs by those in colonial service (district commissioners, district officers, agricultural officers, etc.), I do not doubt that the officers thought they were acting in the best interests of the people and area they served. Nonetheless, the net effect of British policy was to reduce the autonomy of local farmers and to perpetuate a kind of status quo, in which the people and their economy did not develop significantly in the period from 1920 to 1961.

Ujamaa Vijijini

One must realize that a built environment (houses, fields, fences, orchards, roads) is just that, something created by much human labor over a long period of time. The *ujamaa* program, which envisaged larger settlements (agriculturally more productive) in which it became economically feasible to provide important social services, such as schools, health clinics, and piped water, required people to move to newly created villages. I conclude, from the responses to our questionnaire, that the people of Tanga Region did not have their hearts in this enterprise. Although they would list the advantages of larger settlements (piped water, etc.), they were almost unanimous in viewing *ujamaa vijijini* as a calamity, a costly failure, and a human disaster.

The basis for *ujamaa vijijini* (familyhood, or socialism, in the villages) was given in Julius Nyerere's booklet *Socialism and Rural Development* (1967).[6] He stated that to improve the social and economic well-being of rural Tanzanians they needed to follow three principles: (1) they should respect one another, (2) basic goods should be held in common, and (3) all able-bodied people should work. By coming together in socialist villages, it would be possible to provide services such as piped water, a health clinic, and a school; but there were economic benefits to be gained as well. People could cooperate in growing cash crops using appropriate tools and inputs and benefit from economies of scale. These crops could be marketed more effectively through producer cooperatives, with the proceeds used in ways decided on by the people.

The governance of an *ujamaa* village was supposed to be democratic. Each village had an elected chairman, vice-chairman, secretary, and treasurer. They were aided by a village council, with special committees for security, health, and agriculture (Freyhold 1979: 35). There was also a court.

Implementation of *ujamaa* was in its early stages in 1972, when we were doing our first fieldwork. Along the road approaching Handeni from the south (via Mkata and Mazingara, through the famous *Grenzwildnis*, "wilderness frontier"; Giblin 1992: 168), we noticed village building activity. One story told at the time was that President Nyerere traveled that stretch of road to visit these new settlements and encourage the people in their labors. The Handeni district development officer ran the tour. Things went very well at the first village; there was much building activity, excitement, esprit, and cooperative work going on. At the second village, things were not so well in hand, although some construction had started. The presidential delegation was less pleased. At the third village, house lots were only in the first stages of being laid out, and even less progress had been made. The development officer explained that they had only just got started. Given this dismal turn of events, someone suggested that they return to the first village, where there was a thriving *hoteli* (tearoom) where they could get some refreshment. Upon returning to the first village, they found virtually no one there, and much of the construction that had been ongoing appeared to have been dismantled or abandoned. President Nyerere was not amused, and the development officer responsible for these "Potemkin" *ujamaa* villages was dismissed.

The point of the story is that *ujamaa vijijini* was a "top-down" enterprise, a policy of the government that was to be implemented by government officials at district and ward levels. The local people, who had much to lose, were very reluctant to leave their houses and farms. In large degree, village collectivization had to be created forcibly. I am reminded of the time in 1972 when I sat in the office of the district officer in Mkomazi seeking permission to conduct our research while a farmer pleaded for permission to take a truckload of grain to Arusha to sell. His request was denied in a most peremptory, demeaning, and insulting manner. "Top-down," autocratic attitudes, common among district officers and other Europeans in the colonial era, also could be found among government officials after independence.

Planning of *ujamaa* was totally inadequate. Site selection was frequently poor, with questions of water supply and soil quality not answered adequately. The timing of moves was sometimes bad. Sumra (1975a: 96) said that some people in Handeni District were moved to villages after they had planted their crops. This made it necessary for them to clear and plant new fields; they planted late and the crops failed.

One effect of villagization was that it concentrated too many people in nearby fields, which could not bear sustained use without the addition of fertilizer. Some people tried to maintain their former fields, which entailed a time-consuming commute between them and the *ujamaa* village (Sumra 1975a: 101). The communal fields were poorly managed, and farmers gave a higher priority in allocating labor to their own fields. Labor bottlenecks are common, and when farmers encountered one, they would give the labor to the fields they owned, not to the communal plot. This is reflected in yields. Sumra found that yields on *ujamaa* community fields were regularly 33–85 percent lower than on individually owned fields planted with the same crop (the average of 11 fields was 60 percent less) (Sumra 1975a: 105).

In 1993 we asked three questions about *ujamaa:* In your opinion, what were the effects of the *ujamaa* program? Did good things come from the *ujamaa* program? Y__ N__. If yes, what were they? Did bad things come from the *ujamaa* program? Y__ N__. If yes, what were they? It is difficult to know to what extent answers were honest and straightforward or evasive and guarded. In most villages, it was clear that *ujamaa* was disliked intensely; but it may be that some of the respondents regarded us with suspicion. Were we government agents, trying to find out if the *ujamaa* program might be reactivated? What was our real reason for asking such questions? If they said they liked *ujamaa,* would further government help be forthcoming? I have the impression that answers were hedged in Mgera.

In all, some 179 people we interviewed had something to say about *ujamaa vijijini.* Of these, only 38 had exclusively good things to say. Four out of five had some negative comment about *ujamaa* and, in many cases, *only* bad things to say about it. People in five villages (Mzundu, Mwakijembe, Magoma, Vugiri, and Kwamsisi) had many good things to say about *ujamaa*'s benefits; in two villages, they had almost nothing good to say about it (Maranzara and Mkomazi). Respondents in Mkomazi were particularly vehement, since some of them had been shifted to a wild place in Handeni District where several people were eaten by lions. (This also occurred in Kwediamba.) In Mlembule, an early (1972) favorable view of *ujamaa* became mixed, largely because the leaders of the village "ate" all the money.

People were fully conscious of the potentially good aspects of *ujamaa.* They acknowledged both behavioral and attitudinal benefits (cooperation, exchanging ideas, learning from others, working together, and getting to know one's neighbors), as well as material benefits (dispensary, water tap, school, roads, improved transport, free education for all, and ability to market produce more easily). The 179 people who commented on *ujamaa* made 96 comments about the benefits of cooperation and 267 comments to the effect that having a dispensary, school, water tap, new road, or shop improved their lives. There is no doubt that Tanzania's settlement pattern was transformed by *ujamaa vijijini* (fig. 5.23). This did not, however, eventuate in living in a socialist manner, or an economic transformation.

The negative aspects of *ujamaa,* both in the way it was enforced and its social and economic results, far outweighed the positive ones. Further, after the program was officially suspended (effectively after 1979), the infrastructure that had been set in place was not maintained by the government. The provision of water taps is high on the list of benefits mentioned by those interviewed (41 instances), yet in 1993 it took most people as long to fetch water as it had in 1972. In many villages we saw water taps that were no longer functional.

The most striking feature of the responses is the violence (*ukorofi,* "savagery," "brutality") they reveal about how *ujamaa* was implemented. People in virtually all villages, it appears, were forcibly removed from their homes and farms. This was accomplished by political officials, who were aided by armed police and, at times, even military forces (Giblin 1992: 179). Houses were set on fire (over 20 instances) or demolished (table 4.1), sometimes with the "money" and furnishings inside. Still another incentive to join an *ujamaa* village was that in time of food shortage, food relief was given only to those who had joined. Many respondents summarized *ujamaa* and its effects in angry, succinct, pithy language: "There was nothing true inside the program" (Daluni). "There was no reality to it" (Daluni). "*Ujamaa kwa maneno si vitendo*" (Socialism in words, not deeds) (Moa). "We were made poor, we have no *shambas* [fields], life is difficult" (a widow in Magoma). "We were united, wizards, thieves, and good people together" and "Thieves, witchcraft, prostitution, and jealousy have increased" (Kiwanda).

TABLE 4.1. Negative Views about *Ujamaa Vijijini*

Village	House burnt or demolished	Property, fields, crops lost	Stock loss	Theft	Difficulty in shifting to new place	Dishonest leaders	Difficulty with *ujamaa* fields	Other
Mgera	1	1	—	—	2	2	—	
Kwediamba	4	2	3	—	1	—	—	Famine, deaths
Minazini	7	3	3	8	2	1	1	Famine
Kwamsisi	1	6	—	—	2	—	—	
Mzundu	10	12	7	3	2	1	1	
Kwamgwe	4	5	—	1	1	1	—	
Mandera	3	2	—	2	3	1	1	Loss of tradition
Mlembule	2	2	—	—	1	4	1	
Mkomazi	1	2	—	—	4	—	—	People killed
Vugiri	2	6	—	1	7	—	—	
Magoma	5	6	2	1	1	2	—	Famine, loss of tradition
Kisiwani	—	3	—	—	—	—	—	
Kiwanda	—	—	—	2	2	2	—	
Daluni	5	2	2	—	2	2	2	Loss of tradition
Mwakijembe	—	1	4	1	—	—	—	
Maranzara	7	4	—	—	11	1	—	
Moa	1	1	—	1	1	—	—	
Total	53	58	21	20	40	15	6	9

Note: No data available for Kwadundwa.

The economic and social effects were profound. A very large amount of wealth in the form of built structures and improved land (fields, fruit trees, fences, etc.) was destroyed or forcibly abandoned in the move to collectivized villages. Livestock were stolen, lost, fell ill, or died. There was social disruption. There was much theft and mismanagement. "Our children became disrespectful of the parents" (Mwakijembe). Local leaders were humiliated.

People contributed labor to the village communal field. The harvest from the field was supposed to be sold—the profit either divided among people on the basis of their labor contribution or used to buy village improvements (tractor, dispensary). As noted above, these fields were poorly managed, and the profits did not benefit the general population; further, the sites of new villages were often poorly chosen. This happened to people in Mkomazi and was particularly true of those in Magoma, who complained of population pressure and lack of land for fields.

A number of people lamented the loss of traditional places and practices, although no one mentioned abandoning *tandika* sites (places for libation, sacrifice, or other offering to ancestors) or the places where their ancestors were buried as one of the bad features of *ujamaa*. Loss of tradition and revered places, such as graves, is mentioned by Giblin in his discussion of *ujamaa*. It was not that moving from one place to another was unthinkable, but many feared that the leadership of the new villages would be weak and unable to maintain the connection between the people and their ancestors (Giblin 1992: 182).

Governmental force used against local people did not stop with *ujamaa*. The residents of Mkomazi, including Maasai cattle keepers, have had a long history of conflict with wardens and game scouts responsible for the Mkomazi Game Reserve. The game reserve, created in 1951, includes the northwestern part of the Umba Plains in Lushoto District, north of the Usambaras, and lowlands northeast of the Pare Mountains in Pare District. In July 1988, 5,000 residents were forcibly removed from the park by police and game department employees and resettled mainly in villages at the southern and western edge of the park, "where they are disproportionately vulnerable to natural hazards" (Neumann 2001: 313).

Structural Adjustment

The 1980s brought new governmental policies that acknowledged and responded to the serious economic decline and to pressure from the

International Monetary Fund (IMF) and aid donor agencies. Historically, the IMF had been hostile toward *ujamaa*. It had funded only those agricultural projects that involved individual farmers or block farms[7] (Freyhold 1979: 111; Biermann and Wagao 1986). Structural Adjustment Programs (SAPs) were begun unofficially in 1982, and a formal accord with the IMF was signed in 1986. The first phase of structural adjustment was marked by fiscal austerity (Sarris and Brink 1991: 111). The SAPs called for reduction of state expenditures, currency devaluation, and liberalization of imports. Marketing was liberalized and there was greater distribution of inputs for production. Sisal marketing was deregulated after 1983 (Townsend 1998: 259). For farmers, this meant greater availability of equipment, fertilizer, seed, and pesticides. Nevertheless, agricultural production decreased by 0.8 percent annually between 1989 and 1994 (Legum 2000: B411). One result of the SAPs was that the Tanzanian shilling declined in value as an international currency (fig. 4.8). Seven shillings were worth one US dollar at the time of independence (1961). The ratio rose abruptly after 1986.

"[W]ith the onset of the economic crisis and the decline in aid receipts in the 1980s ... the government deliberately placed social services at a lower level of priority" (Doro 1995: B355). The reintroduction of the market economy and harsh economic reforms led to abandonment of agricultural sector programs and severely reduced government expenditures for

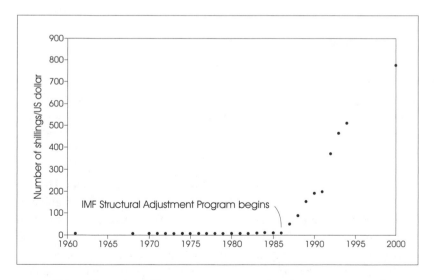

FIG. 4.8. Value of Tanzanian shilling. (Source: Data from *Africa Contemporary Record*, 1970–2000)

research, development, and training (Sosovele 1999: 252). The goal of the
Economic Recovery Program, which ran from 1986 to 1989, was the revival
and revitalization of production (Doro 1995: B355). Government funds to
support education and health were cut sharply. As a cost-sharing measure,
school fees and health fees were reinstated, as well as other user fees. The
government retreated from the socialist goals of *ujamaa vijijini* and relaxed
the leadership code (*mwongozo*), which forbade civil servants and party
members from having outside income (Tripp 1997: 6). The infrastructure
deteriorated as government funds were withdrawn. There were serious
transport problems, both road and rail. In some places, roads were in such
bad condition that some truckers refused to use them (Lofchie 1989: 108).

The economic recovery did not reduce rural poverty or income inequality. In 1989, the Economic and Social Adjustment Program was introduced,
with a goal of rehabilitating basic services, such as transport and the health
sector (Legum 1998: B395). Although Tanga Region registered positive
economic growth from 1984 to 1994, its effects bypassed most of the people. Only about 10 percent of the population was seriously involved in
the market economy. "For the majority of farmers—probably more than
80%—the situation has improved with respect to transport and availability of consumer goods, while incomes in real terms are stagnating or even
declining" (Taube 1989: 9).

The stagnation or slow growth of the economy was evident in Tanga
Town. Once Tanzania's second largest city, Tanga has been surpassed by
Mwanza. Although Tanga grew greatly numerically (27,973 in 1957, 48,134
in 1976, 137,364 in 1988, and 215,432 in 2002), its growth in commerce and
industry was not so swift. Indeed, the central ward (the main downtown)
lost population between 1988 and 2002, going from 7,360 to 6,119 (United
Republic of Tanzania 2003). In 1993 Tanga Town struck us as economically
depressed. There was little traffic on its wide, though poorly maintained,
streets. The large number of empty automobile showrooms indicated that
the car dealership industry had departed. Traffic in the port was down.

As land pressure increased in the Usambara highlands and maize yields
declined, the hill farmers turned to cultivating vegetables as a cash crop
that could be traded elsewhere. Maize does not do well in the highlands,
where it is cooler and the crop takes a long time to mature. "West Usambara farmers can nowadays be happy if they manage to produce 50% of the
maize they need for feeding their families" (Taube 1989: 6). Lowland farmers in Handeni and Korogwe responded by selling surplus maize. Whether
the decline in lowland production of oilseeds and vegetables among the
farmers we interviewed was offset by purchase of vegetables and legumes

in local markets is impossible to say (see chapter 6). The bulk of the vegetable production in the Usambaras is for export to urban markets, mainly to Dar es Salaam but also to Tanga, Arusha, and Mombasa. In any event, many farmers do not live close to a market. Speculation on matters of trade and commerce are complicated by the fact that a great deal of commerce goes unreported, as part of the black market. An estimated 30 percent of economic activity was conducted outside the formal economy (Legum 1998: B396; Maliyamkono and Bagachwa 1989). Sarris and Brink (1991), using different methods, placed it as high as 59–68 percent. By 1989, some 800,000 people in Tanzania were HIV positive (Legum 1998: B389). Many economic writers were pessimistic about Tanzania's performance in the 1980s and its prospects (Collier, Radwan, and Wangwe 1986; Chandrasekhar 1990; Bryceson 1990: 227ff.).

It is clear that the people of Tanga Region would like to be entrepreneurial, but successive administrations have thwarted them, first the Germans, then the British, and again during Nyerere's time as president, during the *ujamaa vijijini* period (1970 to early 1980s). In the era of SAPs (mid-1980s to present), things have loosened up somewhat in Tanga Region. The Regional Planning Office in Tanga recognizes five ecologic zones for Muheza District: (1) cardamom-tea, (2) sisal-maize, (3) intensive fruit culture, (4) cashew-coconut, and (5) extensive grazing (Regional Planning Office 1983: 8). This classification could serve for all other parts of Tanga Region. The cardamom-tea and intensive fruit culture zones (essentially the southwestern quarter of the district) have the greatest potential for development. I mention a number of initiatives in various villages having to do with cash crops, charcoal, planting of orchards, etc., in chapter 5.

Lastly, the reengineered, upgraded road between Chalinze and Tanga (via Segera) was completed about the time we finished our fieldwork. This project, financed and managed by DANIDA (the Danish international aid agency), greatly improved conditions for transport of produce—to Tanga Town itself, as well as all the way to Dar es Salaam. There was a noticeable increase in traffic carrying fruits and other agricultural produce. There also was greater interest in some parts of Tanga Region, especially in the eastern Usambaras (Muheza District), in planting perennial crops, such as oranges, limes, tangerines, pineapples, bananas, and coconuts.

Economic Summary

Early on, it was not clear to the colonial occupiers what sort of economic activity should be fostered. The German colonial authorities viewed the highlands of the Usambaras as an area of great agricultural promise. They promoted coffee, tea, cacao, and teak and noted that all sorts of vegetables and fruit (apples, strawberries) could be grown there. Coffee, with some exceptions (Balangai Coffee Estate), was grown by smallholders. Tea was grown largely on estates. There were experiments with rubber growing: Munro described one major effort made by German and British interests. The biggest development was "in the Tanga-Pangani area where by 1906, some 3,125 acres had been planted with *ceara* (*Manihot glaziorii*) trees" (Munro 1983: 370). This source of latex was viewed as better adapted to the drier conditions of the East African coast than *para* (*Hevea brasiliensis*). There were serious labor problems from the start, and in the end, the higher productivity of rubber from Malaya and other Southeast Asian sources made it uneconomic to continue these plantations. By 1914, the plantations in Mombo, Muheza, and other places along the railway had been abandoned. Sisal (discussed above) was for many years the main commercial crop, and most of its production was in the hands of expatriate owners and managers. In the early 1970s, some 350 cooperative smallholder projects were tried in the sisal-growing areas, but without notable success (Cliffe and Saul 1972: 2.131ff.).

Compulsory markets, established by the British, were abolished in 1960, just before independence. They had not worked well for the local people because of buyer (trader) collusion on prices to be offered. They were replaced by marketing cooperatives; but even these were not satisfactory. Among many problems, the most serious were that the co-op people did not show up on the day arranged for the markets, it was difficult to transport the crop to the market, and the prices offered were too low (Sumra 1975a: 139).

In many respects Tanga's performance in the first decade after independence (1961) was quite good, particularly in the higher-potential areas. In the 1970s, however, Tanga's economy (like that of the country as a whole) stagnated or declined. There were many causes. One was the world oil crisis in the 1970s, which greatly increased fuel costs. Around 1970, sisal began its precipitous decline (fig. 4.7), which meant that opportunities for employment on the estates and for selling food to estate workers were also reduced. There were deteriorating terms of trade (Legum 1981/82: B272). In 1975 there were 208,000 refugees in Tanzania, mainly from Mozambique, Burundi, and Rwanda, a drain on state resources (Legum 1977: B355).

There were ongoing troubles with Uganda. After Idi Amin's coup in 1971, with subsequent expulsion of Asians, there was an abortive invasion of Uganda from Tanzania by a small force loyal to Milton Obote. Amin retaliated with an aerial bombardment of Mwanza and Bukoba. In 1977 the East African Community collapsed, which closed the border with Kenya and Uganda and disrupted the East African Community's many common services. The subsequent war with Idi Amin, when Amin's forces occupied the Kagera Triangle in Bukoba District, cost Tanzania perhaps as much as $500 million, much of it in hard currency (Legum 1988: B449). In 1972 there was an effort to decentralize government, moving decision making away from Dar es Salaam. Also in 1972, the decision was taken not to force socialist living in the *ujamaa* villages; rather, the project became one of villagization. *Ujamaa vijijini* was a major contributor to the poor economic performance during the 1970s and early 1980s because a lot of wealth and productive assets in the form of houses, fields, and trees were abandoned in the move to *ujamaa* villages.

Tanga's people contribute in important ways to Tanzania's economy. The Usambaras produced 17,353 tons of green tea leaf in 1991 (Bureau of Statistics 1992b: 14). Though a ghost of its former self, the region's sisal industry produced 20,076 tons of sisal fibers, 78 percent of Tanzania's production (1992b: 12). Cotton was hardly worth mentioning—48 tons in 1990–1991, or 1 percent of Tanzania's production (1992b: 13). Cashew nuts were important but constituted only 7.5 percent of the national production in 1990–1991 (1992b: 19). Tanga's coastal fisheries contributed 13.9 percent (by value) of the nation's total production in 1989 (1992b: 26).

In general, the local population produces food for consumption within the family and for sale in local markets. Some specialized market garden produce, such as beans and broccoli, is intended for urban markets. There is a lively trade in oranges and tangerines. The Tanga Regional Cooperative Union regularly receives substantial quantities of beans, maize, rice, cardamom, coffee, cashew nuts, and cotton, with lesser amounts of cocoa and sunflower seeds (Bureau of Statistics 1992b: 65). Further details on changes in agriculture and livelihood are provided in chapter 7.

Cultural and Demographic Character

Population

Tanga Region experienced tremendous population growth during the 20th century, and changes in the region's administrative structure reflect responses to that growth. While population may have declined in the first

TABLE 4.2. District Population and Area

District name	Area (km²)	Population 1967	Population 1988	Percent increase	Population 2002	Percent increase
Handeni[a]	13,209	133,235	250,244	87.8	249,572	57.4
Kilindi					144,359	
Korogwe	3,756	140,306	218,849	56.0	261,004	19.3
Lushoto	3,497	210,484	357,492	69.8	419,970	17.5
Pangani	1,425	28,426	37,670	31.7	44,107	17.1
Tanga[b]	4,921	258,609				
Muheza	4,681		229,139	60.1	279,423	21.9
Tanga	240		186,818		243,580	20.4
Total	26,808	771,060	1,280,212	66.0	1,642,015	28.3

Sources: United Republic of Tanzania 1976: 16–17, 1992: 4–17, 2003.
[a]Handeni District was divided into Handeni and Kilindi Districts before the 2002 census.
[b]Tanga District and Tanga municipality were reconfigured in 1975 as Muheza and Tanga Districts.

decades of the 20th century (the influenza pandemic in 1918–1919 affected many in Tanga Region, as did a severe drought in 1925 in Handeni, when 5,000 people left the district [Giblin 1992: 155]), the area subsequently began to recover. Each successive census (1948, 1957, 1967, 1988, and 2002) showed marked growth in population.

Population densities are shown for 1967 in figure 4.9 and for 1988 in figure 4.10; a frequency distribution of the densities of census units is shown in figure 4.11; two age-sex pyramids are provided for 1972 (fig. 4.12) and 1993 (fig. 4.13).[8] A third age-sex pyramid shows absent family members (fig. 4.14). Finally, table 4.2 shows population and areas, table 4.3 shows changes in population densities in the districts for 1967, 1988, and 2002, and table 4.4 shows the amount of land in different population density classes for 1967 and 1988.

The administrative structure of Tanga Region has changed over the years. In 1957 there were five districts: Handeni, Lushoto, Pare, Tanga, and Pangani. By 1967 Pare had been reassigned to Kilimanjaro Region, and as of 1 October 1963, Korogwe District was created from part of Lushoto District (Government Notice 450, cited in Thomas 1967: 26). In the period 1972–1975, decentralization efforts led to the division of Tanga District into Muheza District and a smaller, essentially peri-urban Tanga District. Between the 1988 and 2002 censuses, Handeni District was divided in two, with the western half of the district becoming Kilindi District. The district areas and populations are summarized in table 4.2; the population densities at district levels are given in table 4.3.

TABLE 4.3. Population Density of Districts

District name	Population density[a] 1967	Population density[a] 1988	Population density[a] 2002
Handeni[b]	10.1	18.4	29.8
Korogwe	37.4	58.3	69.5
Lushoto	60.2	101.8	120.1
Muheza[c]	55.5	49.0	59.7
Pangani	19.9	26.4	31.0
Tanga[c]	55.5	778.4	1,014.0
Total	28.8	47.8	61.3

Sources: United Republic of Tanzania 1976: 16–17, 1992: 4–17, 2003.
[a] People per square kilometer.
[b] Includes Kilindi District in 2002.
[c] Muheza and Tanga were part of one district in 1967.

In 1967 a zone of high population density stretched along the coast from Moa, through Tanga Town, to Pangani and in a belt west of the town of Tanga, through Muheza and Korogwe to Mombo, and then northward into Lushoto District (fig. 4.9). North of Muheza, in the eastern Usambaras, was another zone of high density. Almost all densities in the top two classes (over 200 people/km^2 and 80–200 people/km^2) were in these zones. By 1988 zones of high density had expanded greatly (fig. 4.10). There were four areas of continuous densities of over 200/km^2: (1) from the town of Tanga south to Tongoni; (2) north of the town of Tanga, around Mayumboni; and the two large zones in the western Usambaras, (3) one centered on Bumbuli-Soni-Lushoto, and (4) the other on Malindi-Mlalo. When the next two classes (80–200/km^2 and 40–80/km^2) are added, the L-shaped belt of highly populated country is almost continuous.

Some notable patterns of higher density also emerge in Handeni District: a zone of population density of over 200/km^2 lies over Handeni/ Chanika and a zone of between 40 and 200/km^2 is found across east-central Handeni and in western Handeni from Mgera/Kwediboma southwestward. In 1967 the pattern of higher densities (top three categories) was there in an embryonic form, but it did not cover nearly as much area and did not extend south of the Mgera-Chanika axis.

Table 4.4 shows the area in different population densities in 1967 and 1988, together with the change between the two dates. For example, in 1967, 3.1 percent of Tanga Region was covered by census units with densities of over 200/km^2; in 1988 it had increased to 7.0 percent. The top three density

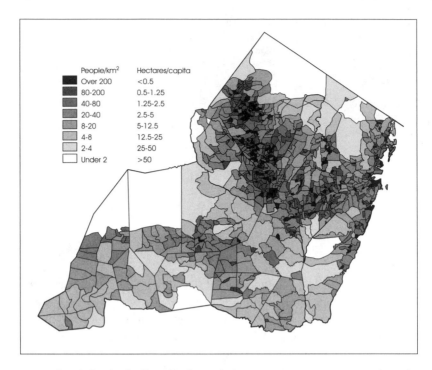

People/km²	Hectares/capita
Over 200	<0.5
80-200	0.5-1.25
40-80	1.25-2.5
20-40	2.5-5
8-20	5-12.5
4-8	12.5-25
2-4	25-50
Under 2	>50

FIG. 4.9. Population density, Tanga Region, 1967. (Source: United Republic of Tanzania 1968)

classes (40–over 200/km²) accounted for 18.7 percent of the area of Tanga Region in 1967 and 35.5 percent of it in 1988, a doubling of territory. The top five categories (above 8/km²) increased in areal coverage. The lowest three categories (below 8/km²) lost area, going from 50.8 percent, or half the region, to 26.4 percent, or a quarter of the region. Study of figures 4.9 and 4.10 in detail will allow one to see how population has grown in different places.

Finally, figure 4.11 is a frequency distribution of the population densities at the ward level. In 1988 there were only 12 wards with densities under 10/km². Most wards had densities ranging from 10 to 155/km², the modal, or most frequent, density being 55/km². Nineteen wards had densities over 600/km² (over 1,500/mi²). These are high densities, indeed; and although most of these wards are urban, some of them are peri-urban or rural. Lushoto and Tanga Districts account for nearly all densities over 600/km² (15 of 19). In many ways, it is a remarkable achievement on the part of Tanga's people that they have accommodated the addition of so many

TABLE 4.4. Amount of Land in Different
Population Density Classes, 1967 and 1988

Density class	1967 Percent	1988 Percent	Percent change 1967–1988
>200	3.1	7.0	3.9
80–200	7.1	11.7	4.6
40–80	8.5	16.8	8.3
20–40	16.9	19.4	2.5
8–20	13.6	18.7	5.1
4–8	21.9	19.4	−2.5
2–4	13.7	3.2	−10.5
<2	15.2	3.8	−11.4

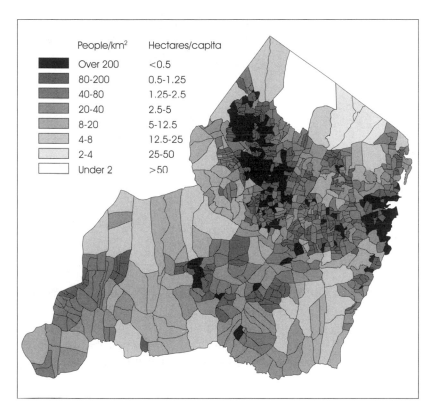

FIG. 4.10. Population density, Tanga Region, 1988. (Source: Data from Bureau of Statistics–Takwimu 1992)

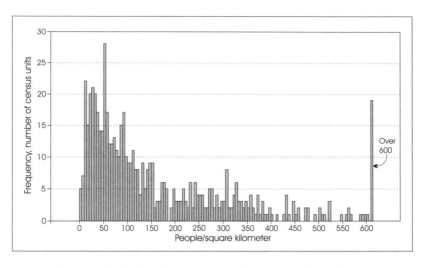

FIG. 4.11. Frequency distribution of population densities, Tanga Region, 1988. (Source: Data from Bureau of Statistics–Takwimu 1992)

people. Agricultural intensification, particularly at higher elevations, has accompanied the growth (Feierman 1993; Turner, Hyden, and Kates 1993: 404).

The two age-sex pyramids of the people we interviewed in lowland Tanga Region in 1972 and 1993 (figs. 4.12 and 4.13) are very similar; but they differ from the conventional age-sex pyramid of a less developed country because they lack a broad base of children in the first two age cohorts (0–4 years and 5–9 years). The modal age for men is 25–29; for women it is 20–24. One should not make too much of the shape of the 1993 pyramid, because it is based on two populations: (1) those in the original sample who were still alive in 1993 (and 21 years older) and who could be interviewed and (2) family members of the person interviewed in 1972, usually a son. The nearly total absence of women aged over 65 is a reflection of two features: the arduous life that women lead and the age differential between husbands and wives (see below).

Ethnicity

Five cultural groups (Zigua, Sambaa, Nguu, Digo, and Segeju) account for 83 percent of those interviewed (table 4.5). People of a wide variety of ethnic origins appeared in our randomly chosen sample of 250 households. Their home areas reflect the history of the sisal industry in Tanga Region,

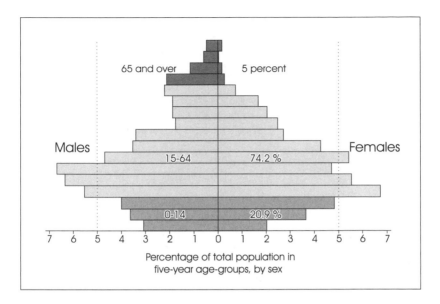

FIG. 4.12. Age-sex pyramid for lowland Tanga Region, 1972 (*n* – 855).

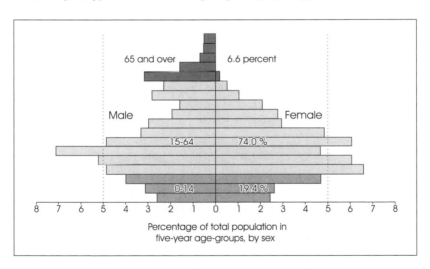

FIG. 4.13. Age-sex pyramid for lowland Tanga Region, 1993 (*n* = 579).

and the most mixed communities (Mkomazi, Magoma, Kwamgwe, and Mandera, which had, respectively, 7, 6, 6, and 4 ethnicities represented in the sample) are also those in closest proximity to sisal estates. The list of ethnic names is a kind of history of labor recruitment and shows the great distances migrants came for work in the sisal fields: from the

TABLE 4.5. Ethnicities of the Eighteen Villages

Village	Zigua	Sambaa	Nguu	Digo	Segeju	Akamba	Taita	Other[a]
Kwadundwa	—	—	14	—	—	—	—	—
Mgera[b]	4	—	10	—	—	—	—	—
Kwediamba	11	—	—	—	—	—	—	—
Minazini	13	—	—	—	—	—	—	—
Kwamsisi	20	—	—	—	—	—	—	—
Mzundu	14	—	—	—	—	—	—	—
Kwamgwe	7	—	—	—	—	—	1	5
Mandera	13	1	1	—	—	—	1	—
Mlembule	11	—	—	—	—	—	1	1
Mkomazi	2	6	—	—	—	—	—	5
Vugiri	1	11	—	—	—	—	—	1
Magoma[c]	—	12	—	—	—	—	—	5
Kisiwani	2	9	—	—	—	—	—	3
Kiwanda	—	7	—	—	—	—	—	4
Daluni	—	2	—	11	3	—	—	—
Mwakijembe	—	1	—	—	—	7	4	—
Maranzara	—	—	—	7	3	—	—	2
Moa	—	—	—	1	11	—	—	2
Total	98	49	25	19	17	7	7	28

[a] Other ethnic groups residing in villages include Arab, Bena, Bondei, German, Hindi, Kilindi, Makonde, Mambwe, Ngoni, Nyakyusa, Nyamwezi, Nyika, Pare, and Zaramo.
[b] One individual was senile and interview was stopped after Questionnaire I.
[c] No ethnic data on one person.

west—Burundi and Nyamwezi; from the south—Ngoni, Makonde, and Nyakyusa; and from the north—Taita. One somewhat anomalous area is Mwakijembe, where most residents are Akamba or Taita, people who settled in the area in the 19th century, leaving what is now part of Kenya. Overall there are 21 ethnicities represented in our sample. The communities with the fewest "outsiders" are Minazini, Kwediamba, Mzundu, and Kwamsisi (all in the heartland of Uzigua), Kwadundwa (entirely Nguu), Vugiri (dominantly Sambaa), and Daluni (almost entirely Digo and Segeju). These latter three villages are all in higher, more remote parts of Tanga Region.

On the other hand, it is inappropriate to make too much of ethnic differences. Kizigua and Kinguu, for example, are mutually intelligible, and among those who speak these languages, the terms themselves refer, respectively, to people of the lowlands and of the highlands (Giblin 1992: 16–18). Similar things could be said of Sambaa and Bondei (Feierman 1974: 72), of Taita and Kamba, and of Digo and Segeju (Gerlach 1965). Sambaa speakers can readily learn the language of the Zigua and Nguu, as well as that of the Bondei (Winans 1962: 7). Gerlach stated that the designation Segeju suggests an individual who is descended of a union in which at least one person had been a slave (Luther Gerlach, pers. comm., 15 December 1999). A history underlies such an assertion. Both Digo and Segeju had slaves until the coming of the Germans, and slavery continued until after World War I (Baker 1949: 36). Slaves had many rights, including the right to marry and to own farms (Kayamba 1947: 96). One form of marriage among the Digo was concubinage with slave women, with the father having "full and lifelong control over his sons" (Freyhold 1979: 17). After 1918, slaves, particularly women, were commonly integrated into the household and family.

The Usambara Mountains have served as a refuge area. One notable refuge group is the Mbugu (a Cushitic-speaking people), who migrated all the way from the Laikipia area of Kenya (Feierman 1974: 75). Since the Mbugu occupy the central and western part of the western Usambaras, far from our transect of 18 villages, I provide no further details on them.

Religion

The dominant religion in Tanga Region is Islam, and it informs all aspects of social life and politics (Nimtz 1980). Evidence of this resides in the names of the people we interviewed. There is such repeated use of common names that at times it leads to confusion. Here are a few lines from the journal my wife and I kept. The entry is for 13 February 1993.

> Toward the end of the afternoon an old man showed up at the katibu's [village secretary's] office asking to be interviewed and claiming to be Abdallah Omari. He even said: "Habari za siku nyingi?" to me. ["What is the news of many days?" as if it had been a long time since we had seen each other.] He had a beaming round face and wore huge glasses that made him look a little like Ray Charles. This happened within minutes of my learning that Athumani Mtibu had interviewed Zuberi, our blind respondent, a couple days ago. Zuberi had not raised the slightest objection to the questions the second time around. This made me wonder if Zuberi was senile and the questionnaire(s) useless. Pitio, when he got a chance later, studied the two sets of answers and they are remarkably consistent. I don't know why Zuberi didn't call a halt the second time around; maybe his life is tedious and even a duplicate distraction is welcome.
>
>   In any event, "Ray Charles" was standing there insisting that he be interviewed. He was good-humoredly sent packing by the katibu, who observed that

this second Abdallah Omari, which apparently really was his name, was usually high on "bhang" at this time of the afternoon. At least we had interviewed the correct Omari, even if we had wasted time in interviewing Zuberi twice.

These are examples of the confusion of names in this Islamicized part of Tanzania. Names often switch back and forth, father to son, so that one generation's Abdallah Mohamed is replaced by Mohamed Abdallah. I made a count of the name frequencies in our list of about 250 people interviewed in 1972. Here are partial results, which group together variants: Athuman, Athumani, Osman, etc.

Name	Occurrence as First Name	as Second Name
Abdalla/Abdallah	13	9
Ali/Alli	9	11
Athuman/Athumani/Osman	4	8
Bakari	8	6
Hamisi	2	3
Hamza	1	4
Hassan	4	2
Hussein/Husseni	5	3
Juma	4	14
Mohamed/Mohammed/Mohammedi	14	11
Musa	1	3
Omari	9	14
Rajabu	2	4
Rhamadan/Rhamadani	1	5
Said/Saidi	6	12
Salim/Salimu/Salum	7	6
Seleman/Selemani	8	1
Total	98	116

These 17 names account for 43 percent of the 498 first and last names among the 249 people interviewed. There were two each of the following: Abdallah Hamza, Ali Abdalla, Ali Athumani, and Rajabu Juma. (Porter and Porter 1993: 72–73)

On the other hand, in the Usambara Mountains, at higher elevations, one finds people with Christian names, a result of the long efforts of Catholic, Anglican, Lutheran, and other denominations. The mix of first names of those interviewed in Vugiri illustrates the point: Jonathan, Charles, John, Justis, Athumani, Saidi, Haji, Ansalem, Ricardo, and Aidan. In Kiwanda and Kisiwani, one finds the following: Julius, Samuel, Timotheo, James, George, John (four times), Michael (twice), Charles, Cyprean, Valentino, and Joseph.

Overall, the communities sampled were dominantly Islamic (Sunni), 10 of them 100 percent followers of Islam. Eighty-one percent of those interviewed cited Islam as their religion, while 19 percent claimed Christianity, usually a branch of Protestantism (Universities Mission to Central Africa, Anglican, Lutheran, or Pentecostal). The following villages were composed entirely of Islamic adherents: Kwediamba, Mzundu, Mgera, Kwadundwa, Maranzara, Moa, Mandera, Mlembule, Daluni, and Kwamsisi. Christians formed a majority in Kiwanda, Kisiwani, and Vugiri. There were significant numbers also in Kwamgwe, Mwakijembe, and Magoma.

Household Size

Average household size was generally 6.5–7, but it ranged among our sample villages from 3.6 in Kisiwani to 9.0 in Minazini, where we did our pilot survey. Our data include others in the household living with the family. The percentage of unrelated people living with families ranged between 2 and 23. Areas with few unrelated people (less than 4 percent) were Minazini, Kwadundwa, Maranzara, Kisiwani, Vugiri, and Daluni. Those with more than 12 percent unrelated people were Moa (the highest, with 21 percent), Mwakijembe, Magoma, Mandera, and Mlembule.

Spousal Age Differential

In 1972 the age differences among husbands and wives in the various villages in Tanga Region were great, with profound consequences for social relations and family dynamics (table 4.6). Middle-aged and older men (especially when taking a second or third wife) commonly married women younger by at least one generation; that is, they married women "young enough to be their daughters." This age differential may have been lessening because the difference in ages between husbands and wives among young married people was much less than among the older people in our study (or perhaps they had not yet reached the age for second or third marriage). There was a difference depending on which generation one belonged to. For men under 30, the wives were, on average, 5 years younger; for men 30–50, their wives were, on average, 10 years younger; and for men 50 and older, their wives were nearly 19 years younger. There was also a difference in a woman's age (relative to her husband) depending on whether she was a first, second, or third wife. First wives averaged 12 years

TABLE 4.6. Age Differential between Husbands and Wives in Tanga Region, 1972

Community	Men's av. age	Husband's age			Differences in age[a]				
		<30	30–50	>50	1st wife	2nd wife	3rd wife	4th wife	5th wife
Handeni									
Kwadundwa	49.2	10.0	13.9	18.7	16.0	20.2	19.5	—	—
Mgera	49.7	2.0	14.4	20.3	16.2	18.0	—	—	—
Kwediamba	52.6	6.0	12.5	19.9	15.0	25.0	—	—	—
Minazini	53.3	6.0	16.0	16.8	14.8	21.0	6.0	—	—
Kwamsisi	41.2	3.8	10.1	24.8	6.3	16.7	12.0	34.0	26.0
Mzundu	46.6	9.0	6.0	16.4	11.4	16.0	—	—	—
Kwamgwe	44.1	6.0	7.1	16.0	10.3	12.5	—	—	—
Korogwe									
Mandera	47.7	4.5	11.6	20.3	13.2	22.0	43.0	—	—
Mlembule	48.1	4.0	8.8	18.2	10.7	22.8	—	—	—
Magoma	47.4	3.0	10.2	18.7	12.8	24.0	25.0	—	—
Mkomazi	41.9	5.0	12.8	20.9	14.8	16.2	19.3	—	—
Vugiri	42.9	3.5	21.7	14.7	11.2	20.0	—	—	—
Tanga									
Kisiwani	46.4	5.0	11.3	13.0	10.9	—	—	—	—
Kiwanda	44.5	3.5	6.8	16.5	9.0	16.7	—	—	—
Daluni	51.1	6.0	11.7	—	8.0	13.7	24.0	—	—
Mwakijembe	50.1	8.3	10.3	20.5	7.4	22.8	34.5	20.0	—
Maranzara	46.7	—	6.0	20.4	14.6	—	—	—	—
Moa	41.6	2.0	9.0	17.0	10.1	—	—	—	—
Means	46.7	4.9	10.4	18.6	11.7	19.0	22.2	28.5	26.0
No. of cases	239	32	126	134	211	54	13	4	1

[a] Years by which women are younger.

younger than their husbands, second wives 19 years younger, and third wives 22 years younger. The average age of the head of household (all men in the randomly chosen sample) was 47 years.

One consequence of the age difference between husband and wife was described to me. It was explained in the following way. Imagine two old men sitting, in conversation, shaded by a mango tree. One suggests to the other: "If you will let me marry your daughter, you can marry my daughter." Faced with such possibilities, young couples may choose to elope, especially when the woman does not wish to marry an older man.

Women experience a great deal of discrimination in their domestic and associational lives. They work harder than men; they have less access to food and to resources; and their economic and political power is more limited. As children, females have a lower priority when it comes to schooling. As adults, they are expected to make good marriages. Useful analyses of women's status and prospects are provided in Mbilinyi 1971; Geiger 1982, 1996; Swantz 1985; Tripp 1994; and Tanzania Gender Networking Programme 1994.

The Proportion of Family Living Elsewhere

Everywhere in Tanzania, one finds people who have moved from the place where they were born. Migration, particularly at the time of marriage and formation of new households, is common (Claeson 1977). In 1972, among the 250 households we interviewed, 290 people were listed as family members who were living elsewhere. Many of these were young men (130) working elsewhere. Some young men of marriageable age left their village because of reduced opportunity to find a spouse in the local area (Feierman 1993: 126). Somewhat more of the absent family members (160) were young married women also working elsewhere, often in the husband's village. Almost no family members over 45 years old lived elsewhere.

The average age of men living elsewhere was 21.4; that of the women was slightly higher, 22.8. There was considerable variation village to village, but one should not make too much of this, for all it takes is one older head of household whose many sons and daughters live elsewhere to raise the overall age of those living away from the village. The range among males was 10.0–35.4, while among females it was 11.4–45.6. The age-sex pyramid in figure 4.14 shows that, for both men and women, those living elsewhere are predominantly in their twenties and early thirties. Migration to other places for work is quite common, as is the sending of money home to the family.

Clientship and Dependencies

Among the cultural groups of Tanga Region there is an embedded tendency toward differentiation, both political and economic. This is certainly true of the Sambaa and the Zigua peoples, but also among the Nguu (Feierman 1974; Giblin 1990; Fleuret 1980; District Books of Tanga Region n.d.: sheet 5, Tanzanian National Archives). In any village one can find people who are wealthier than others. Differentiation is seen in the systems of clientship found among the Zigua and the royal clans and chiefdoms of the Sambaa. For example, among the Zigua, the *si* was instrumental in creating

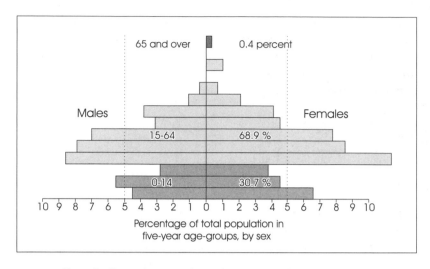

FIG. 4.14. Absent family members: outmigrants by age and sex for 18 villages in Tanga Region, 1972.

large political alliances. The *si* is a small territory (Giblin 1992: 73). It has a meaning similar to *nchi* (Kiswahili), meaning "country," "district," "land," "region." People resident in a *si* honored the ancestors who supposedly founded the *si*. The *si* provided a basis for laying claim to land and also permitted the creation of loyalties based on politics, not simply kinship. The *si,* with its appeal to kinship ties, was used, at times, to contest for political power against unpopular chiefs (Giblin 1992: 76). Some chiefly alliances were substantial in the precolonial period—for example, Rukorongwe (between Handeni and Mombo), Kidunda (between Magamba and Korogwe), and Wakinaluma (between Kiwanda and Muheza) (Giblin 1992: 51). These alliances controlled trade passing through their territories. The alliances all lay along the important trade route that connected Pangani with the interior, passing through such places as Kiwanda, Handeni, Kwediboma, and Mgera. A complex system of trade relations, enslavement, clientship, and patronage was found in these chieftaincies. This system began to decline under German rule and was further disrupted during British colonial times. In 1926 the British introduced indirect rule, which set in place chiefs appointed by the government. The net result was that the system of patronage and reciprocal responsibilities between patron and clients fell into abeyance, with attendant disruptions in trade and control of vegetation and disease (Giblin 1992: 178).

The role of a patron or chief had been to ensure the well-being of the people who owed allegiance to him, as well as of the environment they used. Giblin stated that they "provided farmers with food reserves, seed, arable land and livestock, thus enabling the majority of farmers to remain in their homes, work the land, and maintain a healthy disease environment despite periodic droughts" (1992: 177). Clients, in receipt of the support of a patron, gave up a number of things, the most significant being their autonomy. They had to work as directed, providing labor on projects the patron decided upon. The patron controlled which land they would use, the livestock they would manage, and the location of their houses. Although most would have enough, those with power also had greater wealth.

The accumulation of wealth is illustrated by an excerpt from the journal my wife and I kept in Tanga during our 1993 research.

26 January 1993

Athumani Juma [the name has been changed], reputedly the richest and best farmer in our 1972 survey, seems still to be the richest and best farmer, now experimenting with barbed wire as a means of controlling bushpigs, which if anything are a greater trouble than they were 20 years ago. His house was also a shop and full of old, European-made furniture (turned bed posts, a wonderful cradle, etc., that came from Zanzibar). Pat made an inventory of the items she could see about her....

Pat's observations: Living room 10 × 25, concrete walls, beaver board and wood strip ceiling whitewashed but stained and peeling. Double front door with three vertical panels in each door, ancient with deep (light showing through in places) termite channels, a lovely old Arabic bolt. A two-shelf corner what-not with seashells and cloths on it, a couple of locally made folding chairs and a modern living room set—2 chairs and a couch, wood framed with dark gray plastic seats and red padded arms, grossepointe crocheted antimacassars (ubiquitous in African homes) in bright yellow and red; a Phoenix foot-powered sewing machine, a massive turned wood cradle, like a sawhorse with a new white canvas hammock slung underneath (a grandchild slept in it for a while as I waited); on the walls are hung baskets (decorated or bi-colored rush), ceramic fish, coral fronds and lobster shells, a working pendulum wall clock with a lovely chime, a hanging box/shelf with a smoky bevel-edged mirror with a semicircular top over it, and a corner shelf with a radio and two plants on it. There are more potted plants on the outside windowsill. The window is well-proportioned with two yard-square openings with horizontal bars below (no glass, no curtains) and two smaller openings above with vertical bars. An interior door is covered with an African wild animal print curtain. The dining room with a wooden table and four chairs is a smaller ell extension. (Porter and Porter 1993: 60–61)

Although the Zigua, Sambaa, and Nguu feature elements of political hierarchy, some other groups did not, for example, the Bondei. The origins of the Bondei are ambiguous. Some claim that they are an offshoot of the Zigua (Kiro 1953); others claim that they are a mixture of clans that migrated to the Muheza area from the north (Willis 1992: 205). As an ethnic identity "Bondei" has had a shifting history, flourishing for a time in the mid–19th century, then replaced by another (rather derogatory) term, "Shenzi." The term "Bondei" was revived after 1926 as part of a protest against British colonial policy which granted indirect rule by chiefs to other groups in Tanganyika but denied it to the Bondei (Willis 1992: 203). In any event, this ethnic group has no tradition of royal clans or hierarchical political structure.

Summary and Analysis

We have explored historical, demographic, and cultural features of Tanga as a region. In this final section I will bring together the various elements to see how they form a context in which to understand the farmers of the 18 villages in our transect. The first fact has to do with population growth. Since 1948, population has nearly quadrupled in Tanga Region (461,681 in 1948 to 1,642,015 in 2002). The bulk of the population continues to be rural and gain a living by farming. There has been tremendous expansion of settlement and land clearing and an increase in the amount of land in cultivation (fig. 3.7); but there has been intensification as well (Turner, Hyden, and Kates 1993: 404; Feierman 1993).

Although several different ethnicities are found in the 18 villages, their commonalities are more important than their differences. One can probably tell more about the status of a farm family through knowing its size and composition and the ages of its members than any other feature. Large, well-established families, with access to the labor of their members, are generally richer than smaller, often younger families. The absence of members of a family is frequently a reflection of economic opportunities—the lack of them at home and their existence elsewhere. Islam is the religion of most of the people of the 18 villages. A notable Christian influence was found in three villages—Vugiri, Kiwanda, and Kisiwani, all hill settlements with high environmental potential. There may be differences in worldview among these two groupings, with a greater degree of fatalism that events are God's will among followers of Islam. Polygamy is found in all villages and reflected in striking differences in the relative ages of husbands and wives, particularly in relation to second, third, and fourth wives. The data on family members living elsewhere are linked to spousal age differences, because many young men of marriageable age have reduced opportunity to find a wife in the local area (Feierman 1993: 126). Clientship forms a key theme running through all periods under consideration. There has always been social and economic differentiation among Tanga's people. There have always been rich, powerful, politically connected men, and there have been others—young, old, female, or infirm—who have been poor and dependent, offering their labor to wealthier patrons.

The parade of events in Tanga Region supplies the origins for the social and economic foundation of the region. Because of poor transport, many live in isolation and find it difficult or impossible to move surplus production to markets for sale. The long-standing links of the region with Zanzibar involved an important slave trade and the use of slaves in the local economy. Indeed, slavery can be viewed as another form of clientship, which is such an important feature of economic and social life. The links with Zanzibar are also reflected in the story of trade and commerce in the region— the caravans passing through, the *hongo* (passage fees) exacted from them, and the sale of food to them. The late 1880s brought outright colonial control, first by German entrepreneurs and then by the German government, from which developed a system of governance and administration (the *akidat*) that used educated Zanzibari and coastal Swahili speakers, frequently Arab. Contemporaneous with the arrival of the Germans was the rinderpest epidemic and other disease outbreaks (anthrax, smallpox) that disrupted the relationship among people, livestock, wildlife, and the woodlands and bushlands they used. Early on, the Germans introduced sisal as a lowland crop, plantation grown and requiring the alienation of huge areas of lowland Tanga Region and large numbers of skilled and unskilled laborers. This led to the creation of a large migrant labor force (perhaps as many as 80,000) with high turnover and composed of men coming from far away—Burundi and western and southern Tanganyika. Local people responded by selling food to the sisal estates for their workers. Sometimes local people also worked on the sisal estates. With the rapid collapse of the sisal industry after 1970, the ready market for food supplied to the sisal estates disappeared.

After World War I, a British colonial administration took over from the German one. It was marked by inconsistent policies, but on the whole it discouraged farmers from developing their livelihoods and methods. Such projects and schemes as were undertaken were generally failures and frequently they were unpopular with and resisted by the people. At every turn, obtaining loans and credit, marketing, prices on crops sold, movement of goods, industry, and land use, the British constrained the people and reduced their autonomy.

Within a decade of independence (which came in 1961), the Tanzanian government launched its program of *ujamaa vijijini*. Although laudable in its purpose, it was poorly planned and implemented. It resulted in much economic loss, hardship, and local resentment of government officials. The program was suspended around 1979 and was followed by a period in which economic policy was governed by the requirements of the IMF in the form of SAPs. Although rules on economic activity were liberalized, government social spending—for education, infrastructure, and health—was reduced. There were some hopeful signs in the 1990s for economic development, but many residents of Tanga Region had been bypassed by the economic improvement, and poverty continued to prevail in rural areas.

CHAPTER FIVE

Introducing the Eighteen Villages

The memorable opening sentence of Leo Tolstoy's *Anna Karenina* reads: "All happy families resemble one another, but each unhappy family is unhappy in its own way." One might say the same for the 18 villages in Tanga we studied, although even the several happy villages were different.[1] The main point is that the villages were strikingly different: in the degree to which they were organized to help their members, in the welcome or suspicion the people exhibited at the time of our visit, in the sense of cohesion and dynamism each village showed, as well as in the biophysical endowment of each place.

In what follows I give a brief description of each village's physical and locational setting, including its links with other places in Tanga Region. Next comes a summary of our efforts to reinterview the people interviewed in 1972. In all we were able to interview 113 of those in the original survey, the spouses of 15, and the children of 45. Nine others (a collection of younger brothers, maternal nieces, etc.) made up the rest, for a total of 182 interviews. This is a rate of 78 percent of the original survey (not counting Kwadundwa, which we were not allowed to revisit). The summary of the interviews is followed by a brief description of the rainfall pattern, with a rainfall dispersion diagram for the village. Then comes an evaluation of the village by the student researchers on a number of physical and social dimensions. Comments on the nature of the village and the economy that supports it are given. Changes over a twenty-year period in crops grown are also included. Finally, in most cases, there is a crop calendar for the *masika* and *vuli* rain seasons or for whatever seasons the people of the village use in their farming.

Research Procedure

During this study (in both 1972 and 1993), I did not spend a long time in each community, generally only two to four days in each. Although I was always courteous and respectful, it cannot be said that I established rapport in any way with the people we interviewed. Nonetheless, when I returned 21 years later, people remembered our earlier visit and were pleased that I had come back to find out how their livelihood had changed over the previous two decades.

In 1972 the usual sequence was to arrive at the village in the morning and talk with the *mabalozi ya kumi kumi* (ten-cell leaders)[2] of the village and arrange for a *baraza* (meeting) the next day, at which the sample would be chosen. A full list of the names of heads of household of the community was obtained at this visit, and slips of paper each with the name of a household head and a number were prepared that evening. The next day at the *baraza* the sample was chosen by elders of the community in full view of everyone.

For example, there were 37 households in Kwamsisi Ujamaa Village. We chose 20 respondents, a 54 percent sample (which is about the proportion typically chosen in every village), from a hat containing slips of paper with the names of every head of household. We randomly drew 20 names, plus 4 alternates in the event that an individual was not able or willing to be interviewed. The sequence in Kwamsisi was 6, 26, 34, 15, 7, 18, 5, etc. One individual (34), the third name drawn from the hat, was not available, so he was replaced by the second alternate (32), since the first alternate (28) was "on safari" and could not be reached. Within the framework of the randomly chosen sample, we identified individuals with particular characteristics listed in table 5.1.

Arrangements were made by the students to interview those chosen, some being interviewed that afternoon and others over the course of the next two days. A battery of three questionnaires was administered to everyone, and special questionnaires, each about a particular crop, were administered when it was found that an individual being interviewed grew a particular crop. The general questionnaires covered the following topics:

TABLE 5.1. Farmers with Special Characteristics

Characteristic	Number	Average age
Reputed to be the best farmer	8	49.1
Oldest	11	73.8
Youngest	16	25.7
Mabalozi[a]	28	50.1
Richest	13	49.7
Poorest	8	43.0

[a] Ten-cell leaders.

QUESTIONNAIRE I: LIFE HISTORY, CROPS GROWN, TIMING OF PLANTING, EN-
VIRONMENTAL ASSESSMENT. Besides the topics in the title, this questionnaire also included family characteristics, crop presence, crop importance,
changes in crop use, crop calendars, and drought experience and behavior. There were also notations at the end of every questionnaire about the
respondent's reaction to the interview, the circumstances of the interview,
and the time it took.

QUESTIONNAIRE II: AGRICULTURAL PRACTICES, LABOR. This covered the
mentioned topics, division of labor, use of work parties, and days lost
because of illness. If the farmer grew beans, for example, as a cash crop, a
special questionnaire on beans was also given. There was a special questionnaire for annual crops, one for cassava and other root crops, one for
bananas, and one for tree and cash crops.

QUESTIONNAIRE III: LAND, LIVESTOCK, FOOD SUPPLY, WATER SUPPLY, EX-
PERIENCE AND TRAINING. This covered the farm, its land use, land tenure
and land acquisition, livestock, other sources of food, water supplies and
practices, purchases of agricultural inputs, yields and storage of crop, sales
and purchases, markets, farming experience (short courses and familiarity
with radio and print extension services), plus three attitudinal questions,
including reaction to a brief story about a drought.

Usually only one questionnaire was done at a given time, although the
specialized questionnaires on crops tended to be done when Questionnaire II was being given. Fields were commonly mapped either by another
student, who was shown about by another family member while the interview proceeded, or at the end of the interview.

In 1993, the usual sequence was to meet with the *katibu* (village secretary) and the *mwenyekiti* (village chairman) to determine who among

those we interviewed in 1972 was still living and, if deceased, was survived by a spouse or children whom we could interview (figs. 5.8, 5.11, 5.27, 5.34). We then arranged interviews and in some instances trained others to help with the interviews (Appendix B). We did not remap fields.

Student Researcher Assessments of the 18 Villages

After the 1972 fieldwork had been completed, I had six of the student researchers do an evaluation of all the sites in which they had done interviews (and most had worked in all 18 villages). They were asked to rate, on a scale of 1 to 10, all villages according to a set of measures. They also grouped the villages they felt to be most similar. In the latter task, they were to place the villages into nine groups, and then into four groups. They also stated which crops they believed held the greatest promise for further development in each village. (I also did the rating and grouping exercises.) The students rated each village according to (1) wealth, (2) the degree to which *ujamaa* was successful and supported, (3) the equality of women, (4) the health of the population, (5) soil quality, (6) climatic risk in the livelihood, (7) the potential for good agriculture, (8) the hardship of livelihood, (9) food shortage occurrence, (10) progressiveness (the term was purposefully left undefined), and (11) political consciousness. They also rated each village as to how welcoming and cooperative versus suspicious and uncooperative villagers were in relation to our visit and research. The scores on each item were then converted into rankings. A clear consensus emerged in almost every case as to the character and potential of the village. Not all researchers worked in all villages, so the ratings and rankings are not exhaustive.

The composite rankings are shown in table 5.2. High ranks (i.e., 1 to 5.9) and low ranks (13 or higher) are highlighted as being particularly notable. Some of the rankings are strongly correlated. One example is shown in figure 5.1, which compares wealth with climatic risk.

Students were also asked: "If you had to divide the villages into nine groups, how would you group them? When you place two villages in the same group, do it on the basis that these two are similar and different from other villages." No guidance was given as to what criteria they should consider. It could be climatic risk, wealth, isolation, or simply how the student researchers felt they had been welcomed. In the resulting figure (fig. 5.2), we see that lots of villages got linked to lots of other villages at least once by someone; but there were some notable areas of unanimity. Mkomazi and

TABLE 5.2. Student Researcher Rankings of Villages

Name	Wealth	Ujamaa[a]	Equality of women	Health of pop.	Soil quality	Climatic risk
Kwadundwa	**3.8**	5.0	9.4	7.0	**1.4**	**4.1**
Mgera	8.0		10.6	7.7	**3.5**	**4.9**
Kwediamba	11.2	5.5	10.3	12.7	11.2	*15.0*
Minazini	*16.6*	4.0	**5.7**	*14.1*	12.9	*14.9*
Kwamsisi	*14.3*	5.4	9.3	*13.0*	7.7	*13.4*
Mzundu	12.8	6.6	10.6	*14.3*	9.2	12.2
Kwamgwe	12.9	6.3	11.6	8.2	10.8	11.0
Mandera	8.0	4.8	**5.1**	7.6	7.9	9.0
Mlembule	9.9	**1.0**	7.9	10.8	9.8	11.0
Mkomazi	9.8		11.5	8.7	*16.0*	*13.3*
Vugiri	**3.1**		9.8	**4.8**	7.7	**3.1**
Magoma	9.3	3.4	**4.4**	9.4	9.2	8.9
Kisiwani	**1.3**		9.8	**2.8**	**4.4**	**2.0**
Kiwanda	**5.6**	*8.0*	8.4	**5.5**	7.2	**2.9**
Daluni	6.3		9.4	8.3	9.2	9.1
Mwakijembe	*16.1*		6.2	*14.8*	12.3	*16.4*
Maranzara	11.0	3.3	6.3	7.4	12.0	9.3
Moa	7.2	4.8	9.6	9.4	*14.5*	7.4

Name	Potential for agri.	Livelihood hardship	Occurrence of food shortages	Progressivness	Political consciousness	Cooper./ suspicion
Kwadundwa	**1.7**	*13.1*	**1.9**	9.9	6.9	**5.1**
Mgera	7.1	8.7	8.1	8.3	10.0	8.7
Kwediamba	12.6	12.6	*13.1*	12.3	9.9	**5.9**
Minazini	11.7	*16.0*	*16.0*	*15.5*	11.1	11.5
Kwamsisi	9.0	*15.0*	10.6	11.7	8.4	8.0
Mzundu	*13.6*	12.4	12.0	*15.1*	11.5	7.7
Kwamgwe	11.9	7.4	10.3	9.5	9.7	*13.6*
Mandera	7.6	7.7	11.8	8.6	6.6	6.4
Mlembule	7.3	**5.1**	9.8	**5.5**	**3.4**	7.1
Mkomazi	7.8	7.7	7.7	**5.8**	10.3	11.3
Vugiri	6.6	6.3	**3.3**	**5.4**	12.6	7.4
Magoma	7.3	7.9	8.7	9.7	7.5	**5.6**
Kisiwani	**5.7**	**2.0**	**2.7**	**2.4**	6.3	7.1
Kiwanda	7.9	6.5	**3.6**	7.6	12.4	*13.8*
Daluni	7.2	8.5	6.6	**5.2**	10.3	**1.6**
Mwakijembe	*13.6*	*15.1*	*14.2*	*15.2*	*14.1*	9.4
Maranzara	*13.5*	7.5	11.2	9.9	6.4	*14.9*
Moa	*15.5*	7.0	9.4	6.2	8.2	*13.5*

Note: Highly ranked (1–5.9) in bold; low ranked (13 and higher) in *bold italic*.
[a]Not all villages had experienced *ujamaa* as of 1972.

Mlembule, both dry villages dependent on irrigation, were paired off by more than five observers. Similarly Minazini and Kwediamba, both near Handeni and the first two places where we did research, were strongly paired. Four or more people grouped Vugiri, Kisiwani, and Kwadundwa together. These three villages are in montane forest and have ample

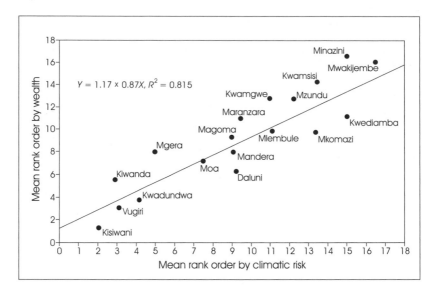

FIG. 5.1. Wealth and climatic risk: evaluations by student researchers, 1972.

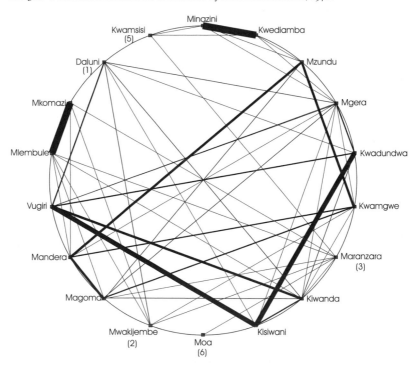

Numbers in parentheses (6) indicate number of non-links with other villages.

FIG. 5.2. Degree of similarity among villages: evaluations by student researchers, 1972.

rain and high agricultural potential. Mandera, Kwamgwe, Mzundu, and Magoma (several of which are strongly associated with sisal estates) were linked by several people. At the other extreme, only one student researcher paired Moa, a fishing village, with another village (one student researcher linked it with Maranzara). Similarly, Kwamsisi, by itself in the far southeastern corner of Handeni District, was seen as separate from the other 17 villages.

Kwadundwa

We did not revisit Kwadundwa in 1993, following the specific advice and firm request of Mrs. Mshakangoto, the Handeni District executive director. She herself, using a four-wheel-drive vehicle, had been stuck eight times in recent days trying to reach the area. We remembered it as the worst road we had been on in 1972. Kwadundwa lies in the southwestern part of Handeni District, 71 km from Handeni Town, in high, rugged, well-watered country. It occupies hillsides stretching out along both sides of the Kwapogora River, between the Nderema and North Nguu Forest Reserves. This forested country, elevation 860 m, is well suited to cultivation of cardamom, coffee, bananas, and other perennial crops. In recent years there has been a lot of illegal timber cutting in the area, with logs going out to the west and north to Arusha and south to Morogoro.

The rainfall of Kwadundwa is unimodal, peaking in April, and averages 1,075 mm/year (fig. 5.3). (The rainfall record stops in 1977.) Drought months are June, July, and August. There is a relatively steady increase in average monthly rainfall from September onward to the April peak. In terms of climatic endowment, Kwadundwa is an area of great potential. Its major drawback lies in its isolation and poor transport links.

The student researchers were unanimous in their view of Kwadundwa. It ranked near the top in terms of agricultural potential and good soil quality. This was an area with low climatic risk and occurrence of food shortages, but it was seen as an area of livelihood hardship, stemming mainly from its extreme isolation and the difficulty of marketing agricultural production. Agricultural vermin were seen as a serious, though controllable, problem, and there was some risk of soil erosion associated with heavy rains. Some felt that were it not for the poor transport, Kwadundwa could progress more rapidly than any of the rest of the villages. In 1993, Kwadundwa was still isolated, its potential not yet realized.

TABLE 5.3. Typical Planting Dates and Growth Periods, Kwadundwa

Rainy season	Crop	Average planting date	Days to maturity	n^a
Masika	Bananas	15 Jan.	330	1
	Beans	1 Apr. (varies)	85	6
	Cardamom	1 May	1.5 yr	2/1
	Cassava	15 Jan. (varies)	350	3
	Maize	6 Feb.	121	8
	Onions	6 June	142	3/2
	Rice	10 Jan.	145	3
	Sweet potatoes	1 May	92	1
	Tobacco	27 May	180	2
Vuli	Beans	10 Dec.	96	2/1
	Cassava	28 Nov.	350	7
	Coffee	17 Feb.	—	2
	Cowpeas	27 Oct.	—	4
	Green grams	6 Nov.	78	1
	Maize	25 Nov.	102	11/9
	Sweet potatoes	11 Jan.	92	1

Note: Characteristic mixes or intercroppings: *masika* rains—maize, cassava, and castor; *vuli* rains—maize, cassava, and beans.
[a] 3/2 means planting date was based on 3 observations and season length was based on 2.

Mgera

Mgera, though seemingly isolated and distant (lying as it does 67 km west of Handeni Town, on the edge of the Maasai Steppe), has a long history of trade and connection with the Swahili coast. Mgera appears in accounts by the Austrian explorer Baumann: "In Mgera everyone wears fine white cloth, muskets and powder are common and almost every child speaks Swahili" (1891: 273, cited in Iliffe 1971: 10; Baumann 1896). There is still evidence of past links with the coast in the characteristic coastal-style, long white garments (*kanzu*) worn by men, their embroidered caps (*kofia*), their names, and the carved doors on some of the older houses. Mgera sits in a north-south col, between the Mgera/Makunguru Hills on the northeast and the Lusanga/Kwamigala Hills to the southwest. The Lukigura River runs through town. The elevation of the settlement is 975 m (3,200 ft.).

In Mgera we had done 14 interviews in 1972. Eight of these people still lived in Mgera, but one was too senile to interview. Three people had died leaving children, two had moved to Gombero, and one had died leaving no kin. We were able to interview nine people in Mgera in 1993.

Mgera has, essentially, one rainy season, which begins in October and peaks in April, followed by a sharp decline in May (fig. 5.4). Four months

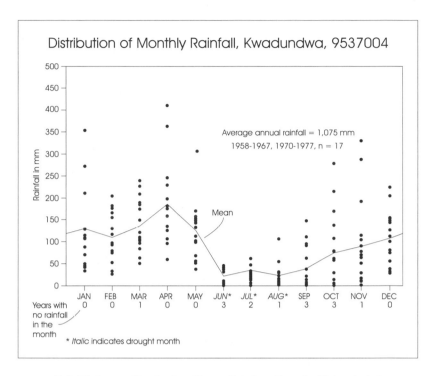

Distribution of Monthly Rainfall, Kwadundwa, 9537004

Average annual rainfall = 1,075 mm
1958-1967, 1970-1977, n = 17

Mean

| Years with no rainfall in the month | JAN 0 | FEB 0 | MAR 1 | APR 0 | MAY 0 | JUN* 3 | JUL* 2 | AUG* 1 | SEP 3 | OCT 3 | NOV 1 | DEC 0 |

* *Italic* indicates drought month

FIG. 5.3. Rainfall diagram, Kwadundwa. (Source: Data from Tanzanian Meteorological Department)

(June–September) are drought months. The average annual rainfall is 821 mm.

This place was ranked highly by student researchers as to the quality of its soils and its low climatic risk. It was average in other respects.

In contrast to many other villages, Mgera has well-developed public facilities such as a modern primary school, a marketplace, village office, courthouse, dispensary, several water taps, government rest house, and well-marked streets. There is a weekly market, frequented by Maasai from the west. There are some modern houses, and many houses are iron-roofed and painted with white lime. Many shops and hotels are found in the town. The village also has at least two maize-milling machines.

Beans, maize, and mangoes are major cash crops (fig. 5.5). Mango trees make a forest surrounding the village. Jackfruit is another important tree crop; each household has at least one jackfruit tree. Bananas (plantains) also can be seen. Livestock is another important economic activity, among

TABLE 5.4. Typical Planting Dates and Growth Periods, Mgera

Rainy season	Crop	Average planting date	Days to maturity	n^a
Masika	Bananas	15 Jan.	—	2
	Beans	25 Apr.	98	7/5
	Castor	15 Jan.	210	3/2
	Cowpeas	1 Jan.	90	2
	Maize	18 Jan.	113	6
	Sweet potatoes	3 May	119	4
	Tobacco	18 May	150	3/1
Vuli	Cassava	6 Dec.	350	6
	Castor	15 Dec.	—	1
	Cowpeas	12 Dec.	89	7/5
	Green grams	25 Nov.	90	5/3
	Maize	15 Dec.	138	7/5
	Sweet potatoes	15 Dec.	171	1

Note: Characteristic mixes or intercroppings: *masika* rains—maize, cassava, castor, and cowpeas; *vuli* rains—maize, cassava, castor, and cowpeas.
[a] 7/5 means planting date was based on 7 observations and season length was based on 5.

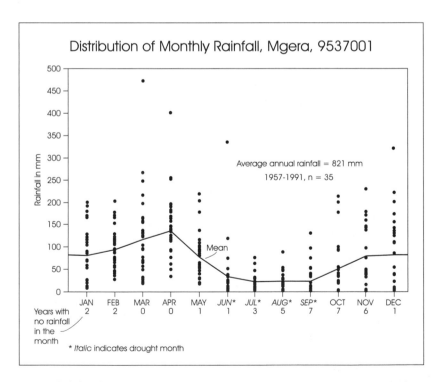

FIG. 5.4. Rainfall diagram, Mgera. (Source: Data from Tanzanian Meteorological Department)

both Nguu- and Maasai-speaking people. One commonly sees cattle herds in the village. There was little change in crops cultivated between 1972 and 1993. A few more farmers increased maize acreage. A couple farmers had given up growing tobacco.

Kwediamba

Kwediamba lies athwart the major road connecting Handeni Town and Chanika with Mgera, at the far western edge of Handeni District. Kwediamba is about 6 km west of Chanika. It occupies sloping, rolling to hilly land that faces east. There is considerable gullying and other evidence of erosion. Settlement is greatly dispersed and no center to the village exists. We met with the village secretary at a partly derelict primary school next to the roadway because the building in which the village chairman had his CCM office had collapsed (Chama cha Mapunduzi, Party of the Revolution, the successor to TANU, Tanganyika African National Union) (figs. 5.6 and 5.7). The village is not named on the topographic maps. There is a cattle dip at the edge of the village, but there are no water taps or shops. Houses are made of mud and wattle construction, with grass thatching for the most part; there are a few iron roofs. There is one mosque, and the names of the residents suggest that most people are adherents of Islam. The scattered houses are often surrounded by bush fencing (such as euphorbia), making house compounds. Livestock are important and some houses have goat and cow sheds. Fields are widely dispersed and some are so far away from the house that at times people live there to guard against bushpigs.

Of the 11 people interviewed in 1972, 4 were still living there (fig. 5.8), 6 had died but were survived by spouse or children, and 1 had died without children. We were able to do eight interviews in 1993.

The rainfall for Kwediamba is represented by the long record made at nearby Handeni (66 years). Rainfall averages 837 mm/year (fig. 5.9). It peaks in April, averaging 164 mm but, like all rainy season months, varies considerably. June through September are drought months.

Kwediamba was seen by the student researchers as a very difficult place in which to live. It has a very unreliable climate, and it experiences frequent food shortages because of failed crops and the destructiveness of bushpigs and other vermin. It does not have good soils or much agricultural potential.

FIG. 5.5. Farm of Zuberi Mnymisi, Mgera, Handeni District. Maize and cassava are interplanted. The maize, planted during the *vuli* rains, is about midway in its season. Photo taken 15 February 1993.

FIG. 5.6. Kwediamba Primary School, Handeni District. This typical classroom has a permanent cement "chalkboard" at the front of the room. Air-dried bricks are stacked: three make a desk, and two, a place to sit. Photo taken 12 February 1993.

FIG. 5.7. Kwediamba Primary School, Handeni District. The "bell" struck to call students to class is a set of wheels from an old sisal estate railway truck. The school has seven teachers and five classes in session at all times. The subjects taught include *siasa* (civics), *hesabu* (mathematics), *sayansi* (science), *sayansi kimu* (domestic science), *jiografia* (geography), Kiswahili, English, and *historia* (history). Photo taken 12 February 1993.

FIG. 5.8. Hamza Ali, age 65, a farmer in Kwediamba, Handeni District, who was interviewed in 1972 and 1993. Photo taken 12 February 1993.

The most important cash crops are maize and seed beans. Also important are sugarcane, cowpeas, and sorghum. Nine different types of cash crops were sold. Maize is overwhelmingly the most important food crop, but also significant are cassava, sorghum, seed beans, and cowpeas. Compared with 21 years earlier, farmers in 1993 were growing more maize and tree crops (such as coconuts, mangoes, and jackfruit). They were growing somewhat fewer vegetables.

Minazini

One reaches Minazini, our original pilot test village, over 7 km of "bad road," southeast of Handeni Town. It is a bad road in a car, but if one is walking, it is a good road. The terrain is rolling and some notable residual hills (e.g., Kiva Hill) lie to the north. It is a relatively short distance to walk to Handeni (the district's administrative headquarters) or nearby Chanika, the commercial center. During the period of *ujamaa vijijini*, many of the people of Minazini moved to Kwedibangala (some 5 km east of Minazini), where an *ujamaa* village had been established in 1970. Since the abandonment of the *ujamaa* policy, some villagers had moved back to be close to their fields, while others had stayed in Kwedibangala.

Of the 13 farmers originally interviewed, we were able to reinterview 8, the children of 2 others, plus an uncle and a niece of 2 others who had died. One person had moved away. We were able to do 12 interviews in Minazini in 1993.

The rainfall for Minazini is represented by the long record made at nearby Handeni (66 years). Rainfall averages 837 mm/year (fig. 5.9). It peaks in April, averaging 164 mm but, like all rainy season months, varies considerably. June through September are drought months.

The main crops of the area are maize and cassava, but one also finds banana groves and fruit trees (including oranges, mangoes, and jackfruit). Some sorghum is grown. Livestock play a significant role in the economy, and a number of villagers have cattle, sheep, goats, and chickens.

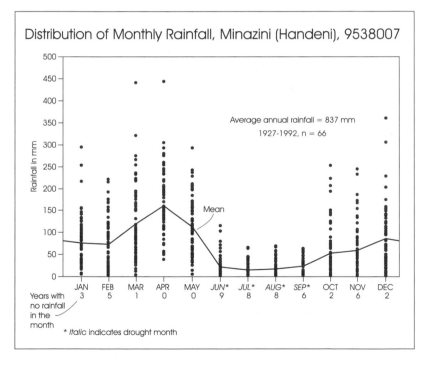

FIG. 5.9. Rainfall diagram, Kwediamba and Minazini. (Source: Data from Tanzanian Meteorological Department)

The student researchers had a strong agreement on Minazini. Like Kwediamba, it was viewed as a poor community, with generally inferior soils and one of the worst places from the standpoint of climatic risk (the average ranking among 18 villages was 14.9). This is a village where food shortages frequently occur and the health of the population is poor. The political consciousness among villagers was viewed as moderate to low, and women were not treated as equals. "Few women are given a chance to attend meetings." It appears that the experience of *ujamaa* in Minazini/Kwedibangala did not have lasting beneficial effects.

The village in 1993 had few improvements. Most houses were of wattle and mud construction and had grass thatched roofs. The only public facility (in Kwedibangala) was a well-constructed three-classroom primary school. The village gives a sense of diffuseness; settlement is widely scattered. The proximity of markets in Chanika has had undoubted effects on Minazini.

There had been two decades of ferment in crops choices, due in part to the shifts in settlement occasioned by *ujamaa*. By 1993 there was much more emphasis on growing maize, cassava, and finger millet. Several farmers had stopped growing leguminous and oil crops, most strikingly peanuts, beans, cowpeas, green grams, and simsim (sesame). This strongly suggests a simplification and possible impoverishment of the diet.

Since Minazini was a test site, some of the questions subsequently asked in other villages were not included. A systematic analysis of crop calendars for food crops and cash crops was not made.

Kwamsisi

Kwamsisi is 36 km from the north-south Chalinze-Segera road, east of Mkata, reached through rugged, attractive *miombo* bushland and savanna. Kwamsisi lies midway between the Chalinze-Segera road and the Indian Ocean and is only 13 km west of where the Dar-Tanga railway line crosses the road, but there is no station there.

In 1972, we did 20 interviews in Kwamsisi. In 1993, 17 of that group were still living in the village; 3 had died but were survived by children. No one had moved elsewhere. We completed 20 interviews in Kwamsisi in 1993.

Kwamsisi has a well-defined bimodal rainfall regime, with strong peaks in April and November (fig. 5.10). The average annual rainfall is 1,013 mm. Only two months (June and August) qualify as drought months. Two distinct crop seasons exist for Kwamsisi farmers. The consequences of poor

TABLE 5.5. Typical Planting Dates and Growth Periods, Kwamsisi

Rainy season	Crop	Average planting date	Days to maturity	n^a
Masika	Beans	6 Apr.	96	3
	Cashew nuts	10 Apr.	3 yr	3
	Cassava	10 Mar.	350	2
	Cotton	10 Mar.	187	6
	Cowpeas	12 Apr.	73	3
	Green grams	20 Apr.	77	4/3
	Maize	12 Mar.	123	18/12
	Rice	20 Feb.	128	2
	Simsim	20 Feb.	169	2/1
	Sorghum	25 Feb.	137	8/6
Vuli	Cassava	3 Oct.	350	7
	Cowpeas	25 Sept.	63	13/8
	Green grams	6 Oct.	70	8/6
	Maize	27 Sept.	107	17/10
	Simsim	1 Oct.	119	10/6

Note: Characteristic mixes or intercroppings: *masika* rains—maize, cowpeas, sorghum, and green grams; *vuli* rains—maize and cowpeas; green grams, cowpeas, and cassava; maize and cassava.

[a] 4/3 means planting date was based on 4 observations and season length was based on 3.

timing in planting are more serious for the *masika* rains, since there is a rapid decline in average precipitation in May and June.

Kwamsisi, though among the poorest villages in 1972, was well organized, cooperative, and extremely hospitable at the time of our 1993 visits (fig. 5.11). The student researchers found Kwamsisi's people poor and in ill health. Livelihood hardship was very high because of climatic risk but also because of poor water supply and quality, lack of transport, agricultural vermin, and trypanosomiasis (this being a tsetse fly area). In 1993, the area was still inaccessible, but the village, nonetheless, had significant local services and appeared to be prospering.

Kwamsisi is a ward headquarters and, despite its isolation, has a well-built office, primary court, cotton storage house, shops and teahouses, a maize mill, water tap, primary school, and dispensary. Maize and cassava are the main food crops, but cotton especially, along with oranges and coconuts, is a significant cash crop. Livestock are important.

From 1972 to 1993 farmers increased significantly the amount of maize and cassava they planted. There was also interest in fruit crops—oranges, pineapples, coconuts, jackfruit, and bananas. Some decline occurred in area devoted to cowpeas, green grams, beans, pigeon peas, and simsim. Six farmers said that they had reduced the area planted to rice.

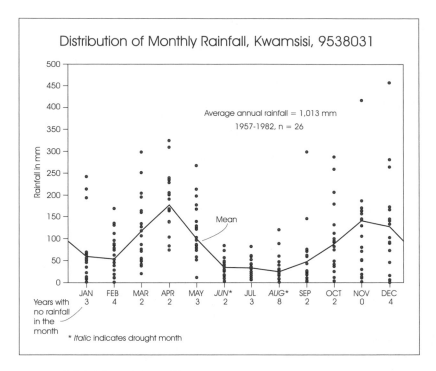

FIG. 5.10. Rainfall diagram, Kwamsisi. (Source: Data from Tanzanian Meteorological Department)

Mzundu

Mzundu lies 13 km west of Kwedikwazu, a settlement on the Chalinze-Segera road. In 1993, it was reached by a very rough, gullied road that would have been impassable after any substantial rain (fig. 4.1). Mzundu sits on the side of a hill, overlooking a market and the headwaters of the Msangasi River. The elevation is 400 m (1,300 ft.). This is quite rugged country, and there is evidence of faulting and drainage obstruction in the area. Immediately east of Mzundu, the Msangasi River flows into a broad swamp, before draining vigorously eastward.

In 1972 we had interviewed 14 people. Of that group, 8 were still living in Mzundu, and we interviewed 6 of them. In addition we interviewed 6 surviving children and the brother of another respondent. We completed 13 interviews in Mzundu in 1993.

Mzundu's average annual rainfall is 845 mm (fig. 5.12). Its distribution is weakly bimodal, with peaks in April (a strong one) and November (a weak one). June, July, and August are drought months.

FIG. 5.11. Four elders in Kwamsisi, Handeni District. The three whose ages are noted were interviewed in both 1972 and 1993. Left to right are Mbezi Ramadhani (65), Mwinjuma Bahari, Husein Abdalla Khalifani (72), and Hatibu Rashidi Ndali (70). Photo taken 19 February 1993.

In 1972 this was one of the least favored villages, ranking near the bottom in the estimation of student researchers with respect to wealth, the health of the population, potential for agriculture, level of climatic risk, and progressiveness. At the time it was very inaccessible. In the 1970s Mzundu was an *ujamaa* village, and it benefited in the provision of public and social services. In 1993 we saw a water tap, a dispensary, a large village office, and a maize mill. There were several shops, a big mosque (made of cement bricks), a church, and a tin-roofed market. It clearly was much more prosperous than it had been. People were well dressed. Although the 13 km road that connects with the Chalinze-Segera road is impassable in the rainy period, for most of the year it can be traversed. There is considerable production and sale of cash crops, which are taken to Kwedikwazu to be sold or taken to urban markets. In addition to cotton, people grow and sell cowpeas, maize, mangoes, and jackfruit. They keep some cattle, goats, and chickens. Several people felt that land disputes were a problem in Mzundu. The major improvement of the Chalinze-Segera road has made a big difference to the people of Mzundu. In the period 1972–1993, the farmers had especially increased their land in maize and cowpeas and had planted more tree crops (oranges, coconuts, bananas, mangoes, and

TABLE 5.6. Typical Planting Dates and Growth Periods, Mzundu

Rainy season	Crop	Average planting date	Days to maturity	n^a
Masika	Beans	1 May	76	7/5
	Cassava	15 Jan.	350	8
	Cotton	3 Feb.	190	3
	Cowpeas	11 Feb.	104	5/4
	Green grams	20 Feb.	133	6/5
	Maize	25 Jan.	125	16/11
	Rice	15 Feb.	120	2/1
	Sorghum	1 Feb.	163	6/4
	Sweet potatoes	1 May	345	2/1
Vuli	Cassava	15 Oct.	350	8
	Cowpeas	25 Oct.	86	8/4
	Green grams	10 Oct.	85	5
	Maize	10 Oct.	141	6/3

Note: Characteristic mixes or intercroppings: *masika* rains—maize, cassava, and cowpeas; *vuli* rains—maize, cassava, beans, and cowpeas.
[a]7/5 means planting date was based on 7 observations and season length was based on 5.

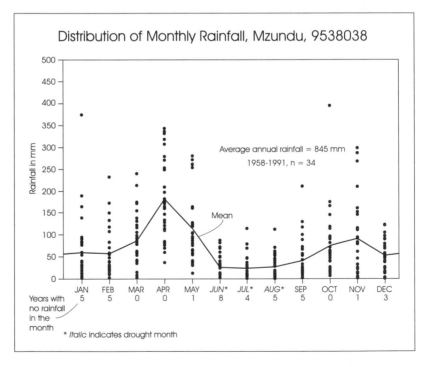

FIG. 5.12. Rainfall diagram, Mzundu. (Source: Data from Tanzanian Meteorological Department)

jackfruit). The major loss has been in cotton; eight farmers had stopped growing it. There had also been some decline in simsim and sorghum cultivation.

Maize is overwhelmingly the most important cash crop, and cassava is a significant cash crop. There is a little trade in cotton, sorghum, and cowpeas. The number of different types of cash crops sold was 12. The inaccessibility of Mzundu and seasonality of crops were brought home to us as we left town. Our field assistant, Pitio, bought about 100 mangoes for almost nothing and piled them in the back of our car. Just outside the town there was an immense pile of rotting mangoes, as if a truck had dumped them there, since they were not worth transporting elsewhere for sale.

Kwamgwe

Kwamgwe is nearly surrounded by Kwamgwe Sisal Estate land. Part of the village is occupied by sisal estate workers' quarters. It is reached from the Segera end of the Chalinze-Segera road at Michungwani ("at the orange trees"), a trip of 10 km. The country is open and rolling, with an elevation of about 300 m.

In Kwamgwe, 6 people of the 13 we had interviewed in 1972 were still living there, 4 had died but were survived by family, 1 had died leaving no heirs, and 2 had moved to Bondo, another village at the end of a long ridge 6 km to the east. We interviewed 9 people in Kwamgwe in 1993.

Kwamgwe has a strongly bimodal rainfall regime, with a peak in April and another in November (fig. 5.13). The average rainfall annually is 1,189 mm. This is a surprisingly high figure given its elevation of about 300 m and its distance from the Indian Ocean. There are no drought months.

Kwamgwe is in the far northeastern corner of Handeni District, far from district headquarters. The people feel a greater affinity with nearby Korogwe and wish they were part of Korogwe District. There were few modern houses, and these were occupied by retired civil servants. There were essentially no public facilities. There was a thatched wooden village office that also served as a classroom. This village stood out in only one respect: the suspicion with which our inquiries were met. It may be that living, as they do, surrounded by sisal estates, they viewed our questions as in some way related to a plan to move them off their land. In all other respects, the village ranked in the middle of the various evaluations the student researchers made.

TABLE 5.7. Typical Planting Dates and Growth Periods, Kwamgwe

Rainy season	Crop	Average planting date	Days to maturity	n^a
Masika	Bananas	5 Feb.	1–2 yr	2
	Beans	15 May	90	3/2
	Cashew nuts	15 Mar.	3 yr	2
	Cassava	10 Mar.	350	10
	Cowpeas	3 May	80	3/2
	Maize	25 Feb.	106	13/9
	Rice	10 Jan.	140	2
Vuli	Cassava	25 Aug.	350	5
	Cowpeas	10 Aug.	90	2
	Maize	27 Sept.	115	8/7
	Rice	1 Dec.	180	1

Note: Characteristic mixes or intercroppings: *masika* rains—maize and cassava; *vuli* rains—maize, cowpeas, and cassava.
[a] 3/2 means planting date was based on 3 observations and season length was based on 2.

The distrust we encountered may have had deep roots. Randal Sadleir, when he was district commissioner of Handeni in the early 1950s, had to settle a dispute between the local people of Kwamgwe and Sir Eldred Hitchcock, the chairman of Bird and Company, the largest sisal estate company in Tanganyika Territory and owner of Kwamgwe Sisal Estate. Sir Eldred wanted to "straighten out" a boundary on the estate, which involved taking a "small sacred grove of trees containing ancestral graves" (Sadleir 1999: 133). Sadleir, despite great pressure from the colonial administration in Dar es Salaam to find for Sir Eldred, decided in favor of the local people.

Cash crops are an important component of agriculture, with emphasis on oranges, jackfruit, and pineapples. Food crops for local consumption are mainly cassava, cowpeas, green grams, and beans. Locally used perennial crops are bananas and mangoes. There are a few livestock, notably chickens. In the two decades since 1973 there was a great increase in cultivation of maize and cassava, as well as the already important fruit crops (oranges, pineapples, and coconuts). Less attention and land area was being given to vegetables and legumes.

Mandera

Mandera is made up of two settlement clusters: Mandera A and Mandera B. Mandera A sits on a hill on the edge of Mandera Sisal Estate. As the

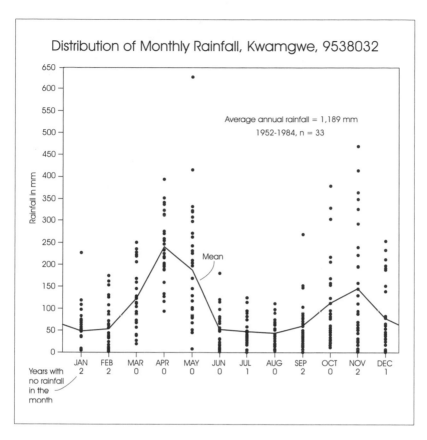

FIG. 5.13. Rainfall diagram, Kwamgwe. (Source: Data from Tanzanian Meteorological Department)

crow flies, Mandera A is 8 km west of Korogwe, but the trip by road is considerably longer. Mandera A has a government office, a restaurant, a bar, and houses that date from German times, including a well-built school, all part of the sisal estate. The *mwenyekiti* (village chairman) lived in Mandera B (or Rutuba), a settlement 2 km to the northeast, near the Tanga Railway line and the Pangani River. It had a dispensary, school, church, and mosque. This is flat to gently rolling country; the general elevation is 380 m.

In 1972 we had interviewed 16 people, but we got information only on the first 13 in the sample. Seven still lived in the village, 2 had died but were survived by a spouse or children, 3 had moved away (1 to Tabora) but we interviewed the daughter of one who had moved away, and 1 was unaccounted for. We completed 10 interviews in 1993.

Mandera's rainfall averages 906 mm annually, the bulk of it coming during the *masika* rains, April peak (fig. 5.14). A slight rise in rainfall occurs during the *vuli* rains, with a November peak. There is only one drought month (September). The tremendous variability in May's rain contrasts with the greater predictability of April's rain.

This old village, set among sisal plantations and close to Korogwe Town, ranked high in equality of women but, in student researcher eyes, was average in all other respects.

Given the close proximity of Mandera to Korogwe Town and to the huge modern Ngombezi, Mandera, and Mawahe Sisal Estates, people have options other than farming. Some have jobs in town or work at times on the sisal estates. There is some growing of vegetables for the Korogwe market. Between 1972 and 1993 maize made major gains, as did rice to a lesser extent. There was a notable decline in cultivation of cassava and beans. Five farmers stopped growing cassava, and nine farmers had reduced the amount of land devoted to beans. At the time of our visit in 1993, people were concerned because their few coconut trees were dying off and they did not know the cause. Seven out of 10 people interviewed referred to land disputes, and a majority of them felt that the problem was severe.

TABLE 5.8. Typical Planting Dates and Growth Periods, Mandera

Rainy season	Crop	Average planting date	Days to maturity	n^a
Masika	Beans	25 May	87	6/6
	Cassava	22 Mar.	350	8/5
	Cowpeas	18 Mar.	87	3
	Green grams	25 Feb.	87	3
	Maize	6 Mar.	134	14/12
	Rice	3 Mar.	126	7/5
	Sweet potatoes	12 Mar.	150	2
	Tomatoes	6 June	63	3
Vuli[b]	Beans	15 Oct.	75	2
	Cassava	3 Oct.	350	5
	Cowpeas	10 Oct.	82	6
	Green grams	1 Oct.	59	4
	Maize	28 Oct.	130	5/3
	Rice	15 Nov.	183	4

Note: Characteristic mixes or intercroppings: *masika* rains—maize, cassava, cowpeas, and green grams; *vuli* rains—maize and cowpeas.
[a] 14/12 means planting date was based on 14 observations and season length was based on 12.
[b] Four people said that the *vuli* rains are not reliable and that they don't plant crops in this season.

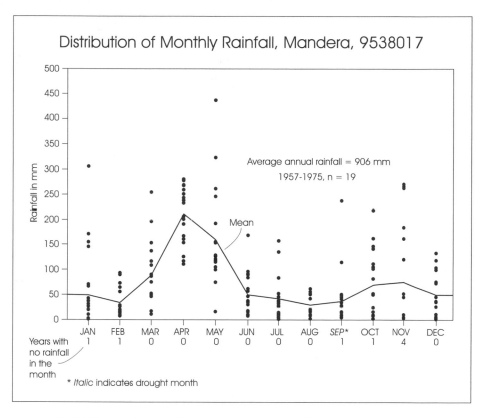

FIG. 5.14. Rainfall diagram, Mandera. (Source: Data from Tanzanian Meteorological Department)

Mlembule

Mlembule sits on the western edge of the floodplain of the Mkomazi River. The channel is braided and flows through large areas of seasonal swamp (fig. 5.15). Some of the villagers' fields are seasonally flooded and some are on higher ground to the southwest. There is very little topographic relief. Mlembule consists of houses set on both sides of a road that connects with Mombo, which lies 5 km to the east. The villagers have planted trees along the road and elsewhere in the village, such as eucalyptus, ornamental acacia, and mango. Attempts to grow coconuts failed. This place is quite windy. Just southwest of the village is a little-used aerodrome that serves Mombo.

TABLE 5.9. Typical Planting Dates and Growth Periods, Mlembule

Rainy season	Crop	Average planting date	Days to maturity	n^a
Masika	Beans	15 Mar.	92	11/10
	Cowpeas	10 Apr.	75	7/5
	Maize	10 Mar.	127	13/8
	Rice	12 Feb.	158	11/10
	Sorghum	25 Feb.	66	2/1

Note: Eight people said that the *vuli* rains are too unreliable to use. Characteristic mixes or intercroppings: *masika* rains—maize and/or sorghum, with cowpeas, green grams, and beans.
[a] 11/10 means planting date was based on 11 observations and season length was based on 10.

In Mlembule, we had done 13 interviews in 1972. Seven of the original group were still living in the village (of whom we interviewed 5), 3 had died but were survived by children (of whom we interviewed 2), 2 had moved away, and 1 could not be accounted for. (The person doing interviews substituted 2 people not in the original sample for those who had moved away.) We interviewed 9 people in Mlembule in 1993.

FIG. 5.15. Mired lorry on the road between Mombo/Jitengeni and Mlembule, across the floodplain of the Mkomazi River, Korogwe District. A tractor has arrived to try to extricate the lorry. Photo taken 2 March 1993.

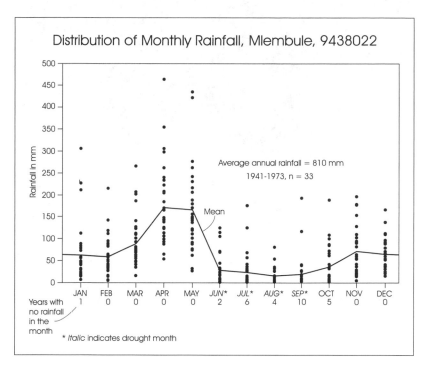

FIG. 5.16. Rainfall diagram, Mlembule. (Source: Data from Tanzanian Meteorological Department)

Mlembule averages 810 mm of rainfall annually, with a strong peak in April and May (fig. 5.16). The months of June through September are drought months. The *vuli* rains are weakly developed. The rainfall in this season is not great, and it varies considerably year to year.

As of 1972, Mlembule apparently had had a very successful experience with *ujamaa*. The *ujamaa* village was started in 1969, and from the first they obtained a tractor and shared a combine with another village. Their water tower was built in 1971. In 1993, however, the *ujamaa* experience was viewed with mixed feelings; some very good things happened, but "the nontrustworthy leaders consumed the money of small people." During our 1993 visit we noticed a high level of esprit and interest in our research (figs. 5.17 and 5.18), but the village itself showed few signs of development. The student researchers rated this place as low in livelihood hardship. It was seen as strong with respect to political consciousness and progressiveness. In other respects, it ranked in the middle of villages.

FIG. 5.17. In every village we visited, youngsters (as well as older people) were fascinated by the topographic maps of their area. They eagerly and readily matched symbols with things in the landscape that they could see: hills, rivers, roads, settlements. Discussions over the maps would go on for hours. These youngsters lived in Mlembule, Korogwe District. Photo taken 2 March 1993.

FIG. 5.18. Mama Idaya Daudi and her daughter, Amina Shatsani, in Mlembule, Korogwe District. Although not interviewed, they were interested in our visit. Photo taken 4 March 1993.

FIG. 5.19. A *hoteli*/bar in Jitengeni, a commercial settlement across the Mkomazi River from Mlembule. The lively mural invites all to come enjoy the camaraderie of this local "Cheers." The Kiswahili is difficult to translate. The man (standing) says: "Welcome, get clean refreshments together with all kinds of drinks." The woman says: "How are things with you? Welcome. Pleasure!" The seated man at the left says: "And truly the friendship is not fierce today." The man at the right chimes in: "Wo, wo, woo, sorry about that." Photo taken 1 March 1993.

In Mlembule there is a maize mill. There is a little-used CCM office, which at the time of our 1993 visit was being used to house pigeons. There is little commercial activity because nearby Jitengeni, a suburb of Mombo on the other side of the floodplain, has many shops and services (fig. 5.19). Rice is the main crop, followed by maize. As for livestock, the people keep mainly sheep and ducks. Fishing is quite important.

By 1993, a little more maize and rice was being grown. Several farmers had stopped growing peanuts and cotton. There was some decline in the cultivation of garden vegetables.

Mkomazi

Mkomazi reminds one of the arid southwestern United States; there is a treeless, sunbaked barrenness and stillness about the place. Some of the people even wear sombrero-like hats. It lies on the east side of the

TABLE 5.10. Typical Planting Dates and Growth Periods, Mkomazi

Rainy season	Crop	Average planting date	Days to maturity	n^a
Masika	Beans	6 July	81	4
	Cowpeas	22 May	60	2/1
	Maize	15 May	146	6/3
	Rice	20 Dec.	184	5/3
Vuli	Cowpeas	1 Nov.	99	1
	Green grams	1 Nov.	99	1
	Maize	20 Oct.	120	2
	Rice	5 Dec.	175	6

Note: Three people said that they plant nothing during the *vuli* rains. The rice is irrigated. Characteristic mixes or intercroppings: *masika* rains—rare, occasionally maize and beans; *vuli* rains—none.

[a] 2/1 means planting date was based on 2 observations and season length was based on 1.

north-south trending Mkomazi River. Irrigated rice fields occupy its swampy floodplain. The houses of the village lie to the east, on a sloping apron of land that rises toward the east, where one can see the bold escarpment of the western Usambara Mountains. Mkomazi has an elevation of 490 m (1,500 ft.). The mountains to the east are over 1,830 m (6,000 ft.). Mkomazi is perhaps more of a town than a village, because the Tanga-Moshi/Arusha rail line passes through the settlement, as does the main road from Tanga to Kilimanjaro, which parallels the railway. (The road and the rail line diverge onto different paths just after one leaves Mkomazi, crossing into Pare District.)

In 1972, we had interviewed 13 people. We found 9 of them still living in the village and were able to interview 6 of them. One had died but was survived by children, 1 had died leaving no family, and 2 had moved away. We were able to complete only 7 interviews in Mkomazi in 1993.

With an average rainfall of only 382 mm/year, Mkomazi is the driest of the 18 villages (fig. 5.20). Eight months (January–February and June–November) are drought months. The rainfall distribution is bimodal; but were it not for irrigation, there would be no agriculture at all in Mkomazi. This area experiences severe droughts and, from time to time, destructive flooding. It is an area of strong winds.

This dusty, semiarid "frontier" town ranked last in soil quality and was among the worst-ranked villages with respect to the severity of climatic risk. Nonetheless, its people were seen by the student researchers as progressive. Though situated at the dry western extremity of Korogwe

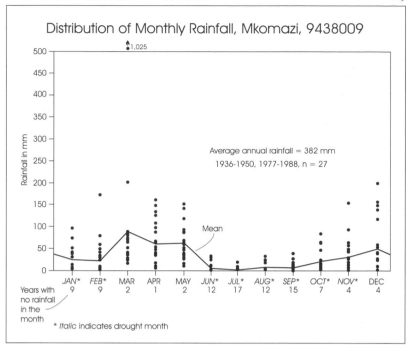

FIG. 5.20. Rainfall diagram, Mkomazi. (Source: Data from Tanzanian Meteorological Department)

District, Mkomazi is well connected by rail and road with the rest of Tanzania.

There are many white-painted, modern, adobe-type houses, and most have tin roofs. Many have water-collecting devices. The town center has several shops, restaurants, a marketplace, and a maize mill. There is a good CCM office, a dispensary/clinic, a courthouse, and a big modern primary school. The streets are well marked. Agriculture is restricted to the irrigated land along the Mkomazi River, where rice is dominant, but some maize is grown, as well as legumes such as cowpeas and green grams. The amount of irrigable land is limited, and several people said that there was severe land pressure and disputes over land. Between 1972 and 1993 farmers in Mkomazi increased the acreage devoted to rice. Three farmers stopped growing cotton. There was a small decline in cultivation of vegetables (beans, tomatoes, onions, and okra). As noted earlier, in chapter 4, Maasai pastoralists in the Mkomazi area had become impoverished and more vulnerable because they had been forcibly removed from the Mkomazi Game Reserve in 1988 (Neumann 2001: 313).

Vugiri

Vugiri is approached by a spectacular mountain road from Korogwe that climbs along the eastern edge of the main Usambara Mountain massif. At one point the road loops back on itself, and by a topographic sleight of hand the ground to the left, which has been a kind of security blanket, suddenly drops away, giving a dramatic view out across the Lwengera Valley to the southeast. This all happens in about 2 seconds as one moves along the mountain road at 25 km/h, and it takes one's breath away—simultaneously magical and unnerving. Vugiri itself looks out across the valley, at an elevation of nearly 1,000 meters. Much of the cultivated land is on steep slopes. The distance to Korogwe by road is about 22 km.

In Vugiri, in 1972, we had interviewed 13 people. On our return visit, we found that 8 of the original group were still living there. Only 3 had died (1 of them survived by children). Two people had moved away. We did 9 interviews in Vugiri.

Vugiri, represented by the long record of nearby Ambangulu Tea Estate, is the wettest of the 18 villages we studied (fig. 5.21). Rainfall averages 2,046 mm/year, with a strong peak over April and May. There are no drought months, and the lower evaporation rates at this highland location mean that it is suitable for many tropical perennial crops, such as tea, coffee, cardamom, and bananas.

Vugiri ranked high in the estimation of the student researchers in several respects, including wealth and health of the people, and low with respect to

TABLE 5.11. Typical Planting Dates and Growth Periods, Vugiri

Rainy season	Crop	Average planting date	Days to maturity	n
Masika	Beans	1 Mar.	90	1
	Cardamom	10 Apr.	1–3 yr	6
	Cassava	1 Mar.	350	9
	Maize	21 Feb.	119	14
	Sweet potatoes	1 Mar.	93?	2
Vuli	Bananas	10 Nov.	1–2 yr	2
	Beans	21 Aug.	100	5
	Cardamom	15 Sept.	1–3 yr	6
	Cassava	15 Sept.	350	6
	Cowpeas	15 Aug.	98	2
	Maize	6 Sept.	129	9
	Sweet potatoes	15 Oct.	155	1

Note: Characteristic mixes or intercroppings: *masika* rains—maize, and cassava; *vuli* rains—bananas, and cardamom.

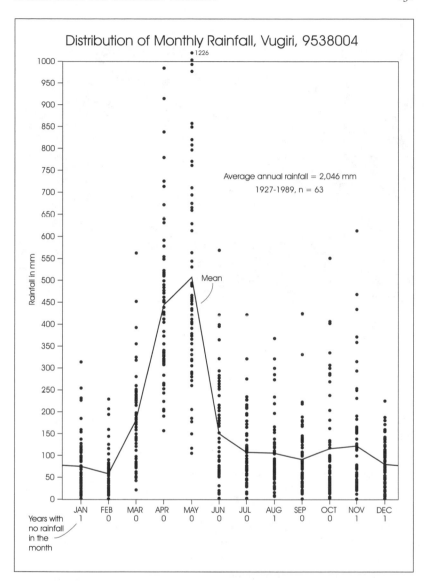

Distribution of Monthly Rainfall, Vugiri, 9538004

Average annual rainfall = 2,046 mm
1927-1989, n = 63

Rainfall in mm

Mean

Years with no rainfall in the month	JAN	FEB	MAR	APR	MAY	JUN	JUL	AUG	SEP	OCT	NOV	DEC
	1	0	0	0	0	0	0	1	0	0	1	1

FIG. 5.21. Rainfall diagram, Vugiri. (Source: Data from Tanzanian Meteorological Department)

the likelihood of food shortages and risk of crop failure. The people were viewed as quite progressive. Most adult villagers had had several years of schooling and were literate.

Vugiri is pleasant and prosperous; tea, cardamom, coffee, maize, bananas, oranges, lemons, and teak are grown (fig. 5.22). Vugiri showed

FIG. 5.22. The houses of Vugiri tightly cluster on a steep hillside at an elevation of about 1,000 m, Korogwe District. Tea is in the foreground. The communal tea plot is at the right in the distance, just below Saint Cyprian Anglican Church. Photo taken 27 February 1993.

the greatest amount of activity and change with respect to crops taken up and abandoned or de-emphasized in the period from 1972 to 1993. Maize and cassava led the way in crop increases. Two farmers stopped growing tobacco, and another reduced the amount planted. There were also reductions in the land devoted to peanuts and garden vegetables.

This is a well-organized village, a ward headquarters. The residents run a communal tea plot for which labor is volunteered, the proceeds of tea sales going to support village projects (dispensary, school, and village administration). There is a strong Christian presence here, centered on Sipriana Mtakatifu (Saint Cyprian Anglican Church) and reflected in the first names of some of its residents. Among those interviewed in 1972 were the following: John, Ricardo, Aidan, Charles, and Jonathan. There is little level land for houses, and fields are on steep slopes, but we saw few signs of erosion. The villagers had even found a way to build a football (soccer) field. (When we met the pastor of Saint Cyprian, he was wearing a Los Angeles Rams practice jersey, no. 40.)

Magoma

Magoma sits among sisal estates near the north end of the Lwengera Valley, a long north-south graben valley that separates the western and eastern Usambara Mountains. The valley is only 3 km wide at Magoma, and hill country rises abruptly to the east and west. In 1993, a good all-weather road connected Magoma with Korogwe, some 38 km, or an hour's drive, to the south. The village is divided by the Lwengera River into two parts: Kawanga (or Kwata) on the west and Mkwaijuni on the east, a newly established settlement independent of the other. Magoma shares the valley with the Magoma Sisal Estates. Houses in Kawanga sit on the east side of the road, set well back from the highway, for there is a large, lively weekly market (Sunday) here that serves a wide region. It is an especially large market for new and used clothing, but the day we were there one could also buy fish, meat, fruit, vegetables, medicines, jewelry, belts, knives, cosmetics, and kitchenware. Magoma grew considerably between 1972 and 1993. The settlement pattern suggests that there was much "villagization" during the time of *ujamaa* (fig. 5.23).

We interviewed 18 people in Magoma in 1972. Only 5 were still living there in 1993. Six had died leaving spouse or children resident; 3 had died leaving no kin; and 4 had moved away (one to Lushoto, one to Dar es Salaam). We were able to complete 11 interviews in 1993.

Magoma's rainfall is only 688 mm/year on average, with one peak in April (fig. 5.24). Drought months occur in two seasons: February and again in July and September. The rains in the *vuli* season (November) are quite low, and crops grown in these rains often fail.

Magoma scored high in two related respects: in the equality of women and its good early experience with *ujamaa vijijini*. (The views on *ujamaa* in 1993 were mixed. Everyone had both something positive and something negative to say about the experience.) The student researchers found the people quite cooperative. In other respects, the village was average.

Magoma is a well-established village, in part because of the aforementioned market but also because it is a division (*tarafa*) headquarters. There are several modern houses, a well-built small market building, several shops, a dispensary, a school, and a good CCM office. The town is served by electricity.

Rice, beans, and maize are important cash and food crops. Seven of 11 farmers had started growing rice since 1972, the only major positive change. A number of farmers had stopped growing or grew less of a wide variety of crops, including cowpeas, green grams, beans, okra, pigeon

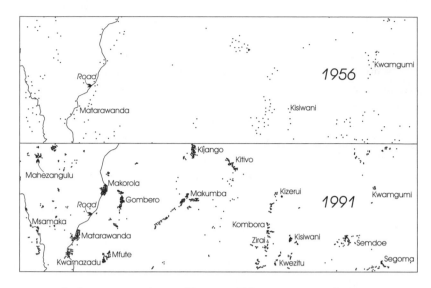

FIG. 5.23. Settlement patterns in part of Lwengera Valley: pre-*ujamaa* and post-*ujamaa*. (Source: Sheet 110/III, 1956, Department of Lands and Surveys, Dar es Salaam; sheet 110/III (Hemagoma), 1991, Surveys and Mapping Division, Ministry of Lands, Dar es Salaam)

peas, peanuts, taros, and sweet potatoes. Livestock and poultry keeping are also important, and people fish in a nearby lake and collect honey. Several people felt that land pressure and land disputes were a severe problem in Magoma.

Kisiwani

Kisiwani, at an elevation of 460 m (1,500 ft.), is situated partway up the mountain road to Amani, between an old teak plantation and the Amani-Zigi Forest Reserve. It occupies a bowl centered on a tributary of the Zigi River. The villagers grow a cornucopia of tropical perennials here: bananas, cardamom, coffee, cassava, coconuts, black pepper, and avocados. Many crops are suited to this cool, moist environment. Although Kiwanda was used as the rainfall station to represent Kisiwani, it is quite likely that the rainfall here is greater and more reliable (fig. 5.25). Kisiwani faces east, the source of rainstorms. Given its higher elevation than Kiwanda's 220 m and the resulting reduction in rates of evapotranspiration, it has a greater agricultural potential. It is so wet here and the rains so reliable that large sections of our questionnaire dealing with drought could be skipped. The

TABLE 5.12. Typical Planting Dates and Growth Periods, Magoma

Rainy season	Crop	Average planting date	Days to maturity	n^a
Masika	Beans	6 May	85	11
	Cassava	6 Mar.	350	7
	Cowpeas	1 Apr.	82	8/6
	Green grams	21 Mar.	75	3/1
	Maize	13 Mar.	127	19/17
	Rice	15 Apr.	189	1
	Sweet potatoes	21 Apr.	140	4
Vuli	Beans	3 Oct.	73	3
	Cassava	15 Oct.	350	5
	Cowpeas	24 Sept.	84	5
	Green grams	1 Nov.	90	1
	Maize	6 Oct.	112	8
	Rice	25 July	153	3

Note: Characteristic mixes or intercroppings: *masika* rains—maize, and beans, maize and cassava, maize and cowpeas; *vuli* rains—cassava and maize.
[a] 8/6 means planting date was based on 8 observations and season length was based on 6.

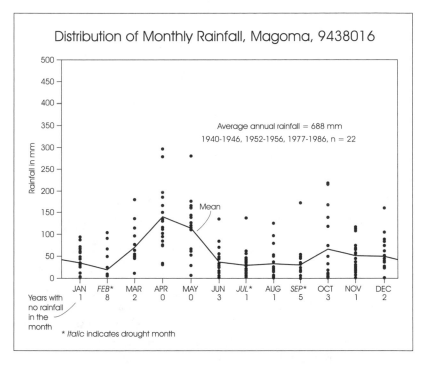

FIG. 5.24. Rainfall diagram, Magoma. (Source: Data from Tanzanian Meteorological Department)

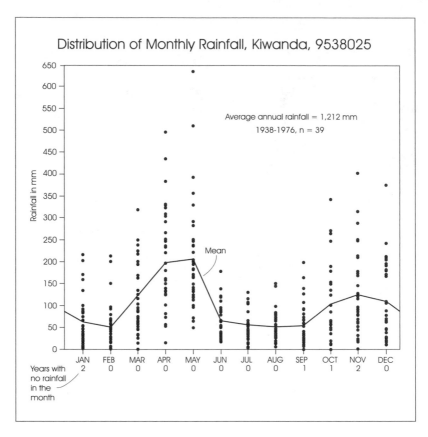

FIG. 5.25. Rainfall diagram, Kisiwani/Kiwanda. (Source: Data from Tanzanian Meteorological Department)

village is scattered but served by a modern CCM building near the road where the village secretary has her office. The distance from Kisiwani to Muheza is 20 km.

In 1972, we had interviewed 14 people in Kisiwani. Only 2 of the original group were still living there, but 2 respondents were located elsewhere. Nine had died, but 6 were survived by family (4 by children, 1 by a spouse, and 1 by a brother) whom we could interview. One had moved away, and another had died leaving no kin. We did 10 interviews in Kisiwani in 1993.

This village is closest to the "Garden of Eden" in student researcher estimation. It must be interesting to live surrounded by cinnamon trees, cardamom plants, clove trees, coffee bushes, tea plantings, avocado trees laden with ripe fruit, and black pepper vines climbing the trunks of huge

FIG. 5.26. Stall feeding of cattle in Kisiwani, Muheza District. This montane area has no problem with drought, and a wide variety of perennial crops, such as coffee, tea, cardamom, cinnamon, avocado, and black pepper, are grown. Photo taken 17 March 1993.

forest trees. In the period 1972–1993, few crops were abandoned, though two farmers gave up growing peanuts. There was a major increase in land devoted to maize and cassava, with lesser increases in bananas and cardamom. In addition to all the perennial and annual crops, one finds stall-fed cattle (fig. 5.26). They are kept in roofed wood stick stables, and fodder is brought to them. One house we visited also had its own fishpond, stocked with tilapia. Kisiwani was ranked first in wealth (there is a thriving cardamom industry, as well as production of many other cash crops), the health of the residents, low climatic risk, and low hardship in livelihood and occurrence of food shortages. The wealth was reflected in the good facilities, which housed the office of the village secretary, the literacy of most of the people we encountered, and the good clothing and jewelry they wore. It was viewed as the most progressive of the 18 villages. The soils are quite good, as is the potential for agriculture. One of the advantages for Kisiwani is resource endowment (good soils and ample, reliable rains), and another has to do with its excellent transport connections with the rest of Tanga Region, since the all-weather road to Amani passes through the village. The road uses the abandoned roadbed of the Tengeri-Sigi railway

TABLE 5.13. Typical Planting Dates and Growth Periods, Kisiwani

Rainy season	Crop	Average planting date	Days to maturity	n^a
Masika	Beans	15 Feb.	105	1
	Cardamom	18 Mar.	1–3 yr	7
	Cassava	10 Mar.	350	10
	Maize	12 Mar.	134	12/8
Vuli	Beans	15 Sept.	60	1
	Cardamom	18 Sept.	1–3 yr	5
	Cassava	20 Sept.	350	8
	Maize	20 Sept.	110	9/7
	Sweet potatoes	15 Aug.	180	1

Note: Characteristic mixes or intercroppings: *masika* rains—maize and cassava, bananas and cardamom; *vuli* rains—maize and cassava, cardamom and bananas.
[a] 12/8 means planting date was based on 12 observations and season length was based on 8.

line, a narrow-gauge railroad built by a private lumber company in the first decade of the 20th century to provide fuel for the wood-burning engines of the Tanga rail line (fig. 3.9). It ran for 23 km, ascending into the eastern Usambaras just past Kisiwani. It was abandoned in 1923. Witness to its existence is preserved in the name of a settlement along its route, Kilometa Saba (Kilometer Seven), as well as the gentle grade of the mountain road itself (Hill 1962: 195; Hoyle 1987: 236).

Cardamom is the most important cash crop in Kisiwani, and during harvesting periods (August–October and January–March) the scent of cardamom is in the air as one drives through the village. In some of the cardamom villages in the eastern Usambaras, growers are Sambaa migrants from the western Usambaras. In Kisiwani, however, most of those interviewed had been born there. Two respondents came from Korogwe District, one from Handeni District, and one from far away Mbeya.

Kiwanda

Kiwanda is a historic settlement, site of a church and mission of the Universities Mission to Central Africa, the group that sponsored David Livingstone. It is reached by a winding road that runs along the east side of a branch of the Zigi River. The road is impassable during the worst months of the rainy season (April and May). Kiwanda's elevation is 220 m, and the village is set in hill country, surrounded by forested lands. The distance from Muheza to Kiwanda is 23 km.

In Kiwanda, in 1972, we had interviewed 11 people. Three were still living in Kiwanda, 3 had moved away, and the other 5 had died. Three were survived by children and 2 by a wife, and we interviewed these 5. One person could not be accounted for. People in this village are widely dispersed and so are their farms. We were able to do 8 interviews in Kiwanda in 1993.

Kiwanda is well favored with rainfall, receiving an average of 1,212 mm/year (fig. 5.25). It is distributed over two seasons, with one peak in May and the other in November. There are no drought months.

Kiwanda was rated significantly high in a number of respects by student researchers: in general wealth of the villagers, the health of the population, absence of climatic risk, and low occurrence of food shortages. However, despite planning (McKay, Daraja, and Mlay 1972), it ranked last as having been successful in its experiment with *ujamaa vijijini*, and the villagers were suspicious of our visits.

In the two decades since 1972 there was a lot of experimentation and change in the crop mix in Kiwanda. Maize and cassava increased greatly in importance, as did coconuts. Four farmers gave up growing cardamom on the grounds that it was too dry there for the crop to do well. A miscellany of garden crops (notably tomatoes, okra, and spinach) were grown somewhat less.

Kiwanda has been the focus of some development efforts. There is a regional training, or folk, college (Chuo cha Maendeleo wa Wananchi) to which students (both young women and men) are brought from elsewhere.

TABLE 5.14. Typical Planting Dates and Growth Periods, Kiwanda

Rainy season	Crop	Average planting date	Days to maturity	n^a
Masika	Bananas	15 Mar.	—	2
	Beans	15 Mar.	60	1
	Cassava	22 Mar.	350	8
	Coconuts	6 Mar.	—	1
	Cowpeas	27 Mar.	80	2
	Maize	18 Mar.	120	9/8
	Rice	10 Mar. (varies)	168	2
Vuli	Bananas	6 Oct.	—	2
	Beans	1 Oct.	62	3
	Cassava	15 Oct.	350	8
	Cowpeas	1 Oct.	76	4
	Maize	6 Oct.	124	6

Note: Characteristic mixes or intercroppings: *masika* rains—maize and cassava; *vuli* rains—maize and cassava.
[a] 9/8 means planting date was based on 9 observations and season length was based on 8.

Topics taught include agriculture, carpentry, and metalworking. The college occupies a hilltop and at times, when the generator is working, has electricity.

Daluni

Daluni sits on lower hillslopes at the northern end of the eastern Usambaras, at an elevation of 400 m. Excellent all-weather roads connect it with Korogwe, through the Lwengera Valley, and with Tanga, via the thriving towns of Maramba and Gombero. The distance to Tanga is 73 km.

We had originally interviewed 16 people in Daluni. Ten of them were still alive and living in Daluni at the time of our 1993 visit and were interviewed. Six others had died, but 4 were survived by children. Two had died leaving no immediate heirs. We did 14 interviews in Daluni (fig. 5.27).

Daluni is well watered, with an average rainfall of 1,102 mm/year (fig. 5.28). Rainfall comes in two clearly delineated peaks, one in April–May and a higher one that peaks in December. There are no drought months, although February averages only 43 mm. The dominance of the November–December peak sets Daluni apart from the other 17 villages and makes the *vuli* rains the main crop season. This distinctive higher peak is probably a result of Daluni's aspect—a northern exposure to the northeast trade winds that develop strongly in November and December. Planting occurs at the beginning of October.

The student researchers who observed Daluni in 1972 had decidedly mixed views about the place. Some thought the hardship of livelihood was great, that the population was in poor health, and that food shortages were to be expected. Others gave it higher ratings. They all gave it middling ratings as to quality of soils and potential for agriculture, but they felt that the people were progressive in outlook. They found them the most welcoming and cooperative of the villagers we interviewed. The student researchers would have been surprised and gratified to see the development in agriculture and in social and economic services that have occurred in two decades.

Daluni has grown a great deal in two decades, and it shows major signs of being connected with the rest of Tanzania. While we were there one day, a Toyota land cruiser arrived, a truck, two Moud busses, the bus of another company (Muanjala), and a motorbike. There is regular bus service to

FIG. 5.27. A rather public interview with Salimu Abdallah, Daluni, Muheza District. Pitio Ndyeshumba records Salimu's responses, while youngsters clown for the camera. Photo taken 31 January 1993.

Tanga at least twice a day, and perhaps more frequently. There are lots of bicycles, and youngsters riding them. We saw up to five kids on a bike. Of course, bikes are also used to carry goods, such as large hampers of oranges and bags of charcoal. On the road to Tanga we encountered groups of cyclists. Those going toward Tanga were loaded and had an ungainly appearance, each with a big bag of charcoal sticking out on either side at right angles to the bicycle. If the bike riders were going in the other direction, one often saw on the back of the bike a "furled" jute bag, wrapped around the stabilizing sticks and secured by the rubber strips or rope used to tie the charcoal.

Daluni is an administrative center with a CCM branch office, ward offices, and a primary courthouse. There are shops, coffeehouses, hotels, and a marketplace. The buildings along the wide main street are tin roofed. Elsewhere most houses are grouped in compounds, made of mud, wattle, and grass thatching and surrounded by coconut, jackfruit, and orange trees. This is an area of great agricultural potential. Farmers grow cashew, rice, maize, cassava, coconuts, and, up in the hills, cardamom. We also noted irrigated taros in some of the lower areas adjacent to streams. In the period

TABLE 5.15. Typical Planting Dates and Growth Periods, Daluni

Rainy season	Crop	Average planting date	Days to maturity	n^a
Masika	Beans	12 Apr.	82	4/1
	Cashew nuts	10 Apr.	5 yr	4/1
	Cassava	27 Mar.	350	9
	Cowpeas	1 Apr.	77	7/5
	Green grams	6 Apr.	77	3/2
	Maize	1 Apr.	114	14/8
Vuli	Bananas	1 Oct.	1–2 yr	2
	Cashew nuts	25 Sept.	5 yr	3
	Cassava	3 Oct.	350	12
	Cowpeas	3 Oct.	93	7
	Green grams	3 Oct.	119	3
	Maize	6 Oct.	112	8

Note: Characteristic mixes or intercroppings: *masika* rains—maize, cassava, and cowpeas; bananas, maize, and cassava; *vuli* rains—maize and cassava.

[a] 4/1 means planting date was based on 4 observations and season length was based on 1.

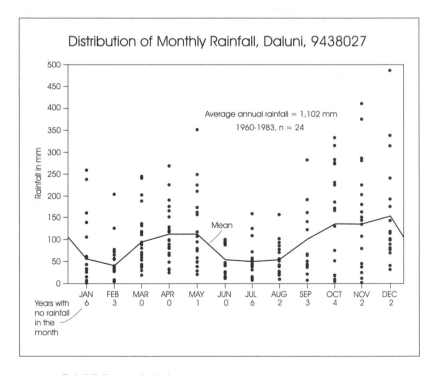

FIG. 5.28. Rainfall diagram, Daluni. (Source: Data from Tanzanian Meteorological Department)

1972–1993 there was a great increase in cultivation of cassava and maize. On balance, there was a small increase in cultivation of perennial fruits, notably coconuts and pawpaws (papayas) but a decline in the area devoted to beans, taros, green grams, and cowpeas. There are few livestock. People said there is a very serious problem with bushpigs; six of those interviewed said that land disputes were also a problem.

Mwakijembe

Mwakijembe is one of the more remote, inaccessible of the 18 villages, lying on gently rolling lowlands adjacent to and south of the Umba River, in the northern parts of Muheza District, only 6 km from the Kenya border. It is essentially unreachable from the main north-south road connecting Tanga with Mombasa and is reached rather from Daluni, which lies 40 km to its south. The river is flood prone and at the time of our 1993 visit had destroyed many maize fields that had been planted near the river. The crocodiles were coming out of the river into town. Across the river is the Umba River Game Controlled Area. In the mid-1950s, settlement occurred on both sides of the Umba River. Indeed, the name Mwakijembe was shown in the area north of the river on older maps. The village has few public facilities and lacks water taps, shops, and maize mills. There are churches and mosques and a village office but no modern houses. The dominant cultural-linguistic group in Mwakijembe is Akamba, many of them from Kenya, but the community is quite ethnically mixed. Livestock keeping is a very important activity here, and some people have large herds of cattle as well as large numbers of goats. Hunting is also an important part of livelihood.

In 1993 we interviewed 6 of the 12 people we had interviewed in 1972, and children, brothers, or surviving spouses of most of the others, for a total of 11 interviews. In establishing who was still living, I read each name to the *katibu* (village secretary) and *mwenyekiti* (village chairman). As I hesitatingly read the name of the oldest farmer interviewed in 1972, they both laughed, since he was still living, aged 94.

Mwakijembe is the most climatically risky village in this study. Although Mkomazi averages less rainfall (382 mm), the people there depend on the Mkomazi River for irrigation and thus are not directly subjected to the vagaries of variable and unreliable rainfall. Mwakijembe averages only 485

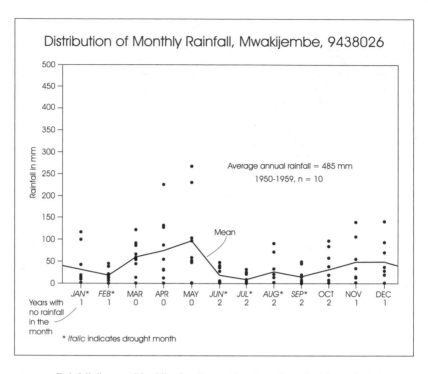

FIG. 5.29. Rainfall diagram, Mwakijembe. (Source: Data from Tanzanian Meteorological Department)

mm of precipitation annually, the bulk of it falling in April and May, with a smaller peak in November and December (fig. 5.29). January and February are drought months, as are the four months from June to September. Quick-growing, drought-tolerant, and drought-escaping crops are needed in this area.

The students who interviewed people in Mwakijembe generally agreed on the kind of place it was. Overall, it ranked at the bottom of the 18 villages. It was climatically risky, with poor agricultural potential and frequent food shortages. Livelihood there involved much hardship and the people were poor and in poor health. Mwakijembe was not seen as a progressive village, and the political consciousness of the people was low.

The village had grown considerably from the earlier visit. Aside from the road from the Daluni-Maramba turnoff, which is improved from 1972, we saw little development in the town. There are lots of cattle; there are large maize fields, especially on the floodplain of the Umba River. In the two decades since our 1972 visit, farmers were growing more maize, but

FIG. 5.30. Oranges for sale—custom peeled if you wish—at outdoor stand, Mwakijembe, Muheza District. Photo taken 30 January 1993.

there had been considerable reductions in other crops. Eight of 11 farmers interviewed had given up growing cassava, and there was a steep decline in cultivation of green grams, cowpeas, and finger millet. We saw no shops, and only one outdoor stand selling oranges (fig. 5.30). The area to the south of Mwakijembe is lightly infested with tsetse fly. It probably has water problems as well during the dry season. The economic emphasis in Mwakijembe is on livestock and charcoal making, with its risky agriculture seen as a necessary, but less important, part of livelihood. The government tried to create a cattle market in Mwakijembe but the cattle owners refused to cooperate.

There is some mining and prospecting for tanzanite and other precious and semiprecious stones, such as rubies, sapphires, and tourmaline, found in surficial alluvium. In the mid-1980s, 51 prospecting licenses had been issued, and Umba Mines, a foreign-owned company, employed 200 people. Because of the proximity to the Kenyan border, many of the stones are smuggled out of Tanzania (Regional Planning Office 1983: 23–24).

There is a great deal of charcoal making in this bushland/woodland both to the south and to the east. We encountered trucks loaded with charcoal in the area. The road we took from Duga (on the Tanga-Mombasa road, north of Moa) to Mwakijembe had tracks going all over the place, some of

TABLE 5.16. Typical Planting Dates and Growth Periods, Mwakijembe

Rainy season	Crop	Average planting date	Days to maturity	n^a
Masika	Cassava	1 June	350	3
	Cowpeas	3 May	53	6/5
	Green grams	3 May	53	5/4
	Maize	25 Apr.	107	14/10
	Sorghum	1 July	69	1
	Sweet potatoes	20 May	120	2
Vuli	Cassava	10 Oct.	350	3
	Cowpeas	10 Oct.	77	3/4
	Green grams	10 Oct.	76	3/3
	Maize	28 Sept.	116	5
	Sorghum	1 Nov.	129	1
	Sweet potatoes	20 Oct.	99	2

Note: Characteristic mixes or intercroppings: *masika* rains—maize, cassava, cowpeas, and green grams; *vuli* rains—maize, cassava, cowpeas, and green grams.
[a] 6/5 means planting date was based on 6 observations and season length was based on 5.

which we took and which landed us in the center of burned places where people had fired charcoal. Again, leading off the road from Mwakijembe to the Daluni-Maramba road were truck tracks, some well developed, that went to charcoal-making areas. During the several days we were doing our work in Mwakijembe and Daluni, there was a lorry loaded with charcoal that had broken down on the big hill just north of Maramba. One day as we returned home, we saw three other charcoal-loaded trucks stopped near the one that was broken down. Later that night they all roared through town (Maramba) past our bedroom window.

Maranzara (Pongwe)

The village we actually worked in is called Maranzara (it means "finish or end hunger" in Kidigo) and lies 2.5 km south of the town of Pongwe, which itself is on the Tanga-Muheza road. This is flat to gently rolling country composed of old marine terraces and relic dunes. The elevation is 80 m and the soils are exceedingly sandy. The village lies only 5 km southwest of a cement factory, whose stack with its noxious plume can be seen in the distance. Many people in this area suffer respiratory illnesses because the cement plant has no filters and the plume often persists for long periods over the countryside.

TABLE 5.17. Typical Planting Dates and Growth Periods, Maranzara

Rainy season	Crop	Average planting date	Days to maturity	n^a
Masika	Cassava	6 Mar.	350	4
	Cowpeas	1 Mar.	82	7/4
	Green grams	1 Mar.	75	5/4
	Maize	15 Mar.	151	9/7
	Rice	20 Apr.	125	3/2
	Simsim	15 Mar.	141	1
	Sorghum	15 Mar.	131	2
	Sweet potatoes	15 Mar.	141	2/1
Vuli	Cassava	20 Sept.	350	5
	Cowpeas	20 Sept.	71	7/6
	Green grams	15 Sept.	74	6/5
	Maize	20 Sept.	139	3
	Simsim	15 Sept.	141	1

Note: Characteristic mixes or intercroppings: masika rains—maize, cassava, cowpeas, and green grams; vuli rains—maize, cassava, cowpeas, and green grams.
[a] 7/4 means planting date was based on 7 observations and season length was based on 4.

In Maranzara, 6 people from among the group of 12 we had interviewed in 1972 were still living there and were interviewed; 6 had died. We interviewed 2 surviving widows and 3 children. One had died leaving no kin. We interviewed 11 people in Maranzara in 1993.

This is a well-watered village, averaging 1,163 mm/year (fig. 5.31). There is a strong peak in rainfall in April and May and a weaker one in November, but there are no drought months. There is considerable variability in the amount of rain that comes during the rainy periods. Years of severe drought are to be expected.

Maranzara ranked in the middle of villages in most respects but was seen as being notably low in agricultural potential, largely because of its poor, sandy soils. The ujamaa village experiment initially had been a success. The student researchers rated the villagers as highly suspicious of our presence, the most suspicious of all the villages.

The village occupies both sides of a north-south road, along which are public and commercial buildings. Poles carrying electricity also line the road. Houses are scattered about behind this line of settlement. There is a well-built primary court with offices for the village secretary, a substantial multiroom school built of whitewashed concrete blocks, with three painted maps on the outside walls, one of Tanzania, one of Africa, and one of the world (using a Mercator projection). There is a borehole with an associated round water tank, but it did not appear to be in use.

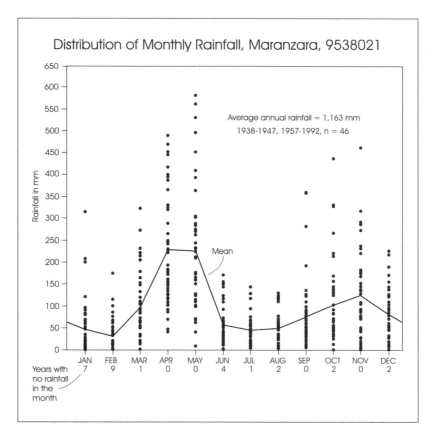

FIG. 5.31. Rainfall diagram, Maranzara. (Source: Data from Tanzanian Meteorological Department)

Agriculture in Maranzara is based mainly on maize, cassava, cowpeas, and green grams, with much interplanting of pulses with maize and cassava. Pigeon peas are cultivated, and we saw some that were old and had become small trees. Crops are widely spaced in the field because of the sandy soils. Over the period from 1972 to 1993, there was quite a lot of shifting in crop emphasis, most notably in maize and cassava, which increased. Fruit trees had also been given greater emphasis. Declines were registered in cultivation of vegetables (tomatoes, okra, sweet potatoes, simsim, green grams, and pigeon peas). Six of 11 farmers stopped growing finger millet. We were shown a starchy potato-like tuber called *wanga* (*Maranta arundinacea*, arrowroot) that is used for ironing and also as a famine food. Local Arabs mix it with oil, groundnuts, and sugar to make *halua*, a Turkish sweetmeat. Other crops we noticed during our visits were okra, sesame, and an orchard

FIG. 5.32. Houses in Moa, Muheza District. Roofing is made of preassembled palm branch shingles. Photo taken 26 January 1993.

of tangerine trees. Many of those interviewed felt that land disputes were a moderately serious problem.

Moa

Moa is a thriving fishing village. It is situated on the Indian Ocean, sheltered by Moa Bay on a bit of shore that is free of fringing mangrove swamps. It lies 2 km off the coast road that connects Tanga with Mombasa, about 50 km north of Tanga. Moa was once an important Arab settlement, with considerable dhow traffic (fig. 5.32). With the suppression of the slave trade by the Germans, it "reverted to the status of a fishing village" (Freyhold 1979: 17). The derelict remains of a long jetty run into the ocean (fig. 5.33). Although this suggests that fishing used to be more important, the fishing industry is still vigorous. Boats simply come to the beach now. There is a lively fish market twice a day. Trucks arrive from time to time to transport fish, and there is a regular bus service. It serves a wide area, and fishermen from nearby villages, such as Kiriju (to the north) and Peluhiza and Mwagoza (to the south) also bring their fish here. Indeed, boats from Pemba and the Kenya coast (Vanga and Kwale) and Monga visit Moa to sell fish. It

FIG. 5.33. Abandoned jetty at Moa, Muheza District. Moa, an old Arab settlement with much dhow trade, declined with the suppression of the slave trade. It is still an important fishing village, with two markets a day. Photo taken 26 January 1993.

is claimed that the prawn catch is smuggled to Kenya (Regional Planning Office 1983: 20).

Normally the people set out to sea between three and four a.m. and return between one and four p.m., but they may return as late as 11 p.m. if they encounter adverse winds. In 1971, the town had 60 *ngalawa* (a small dugout canoe with outriggers) and two boats (belonging to the Fisheries Department). A canoe takes a maximum crew of 6. There were also three small dhows, taking a crew of 12–15. The dhows can fetch 1–2 tons of fish from the sea per day, whereas one fisherman on a canoe can fetch 6 *korija*, partly because canoes can fish only within the reef. Since a *korija* is a measure of 20, this translates into 120 fish. Some fishing is done with dynamite and with hand grenades, especially in the inshore area, inside the

FIG. 5.34. Muajamaa Jumaa, his wife, Fatuma, and some of their children, in Moa, Muheza District. He was interviewed in both 1972 and 1993. Photo taken 26 January 1993.

reef. For all of Moa (including tributary villages) there were 427 fishermen and 119 fishing vessels. In 1971–1972 they netted 785.4 tons of fish, having a value of T Sh 1,103,700/- (about $131,000 in 1972 US dollars). The average price was T Sh 1/35 per kilo ($0.16). Farming in Moa is a sideline.

Seven of the 14 people we interviewed in 1972 had died. By talking with surviving spouses or children, we were able to do 11 interviews in 1993 (fig. 5.34). One of the farmers interviewed during our earlier visit, reputedly the richest and best farmer in the village, still seemed to be the wealthiest as well as the best. He was currently experimenting with barbed wire as a means of controlling bushpigs, which, if anything, were a greater trouble in 1993 than they had been 20 years before. After fishing, people are interested in cashew and coconuts as a source of cash income. Little maize is grown; cassava has greater prominence. There is also salt making along the coast. Salt cooking requires a fire and a lot of firewood, and this has led to great devastation in nearby forests (Regional Planning Office 1983: 22).

The observable differences in wealth in Moa can be traced to the way the fishing industry is organized. Much of the infrastructure of fishing is controlled by small capitalists who rent boats and nets to fishermen. The price of rental is half the catch. The fishermen are further at a disadvantage

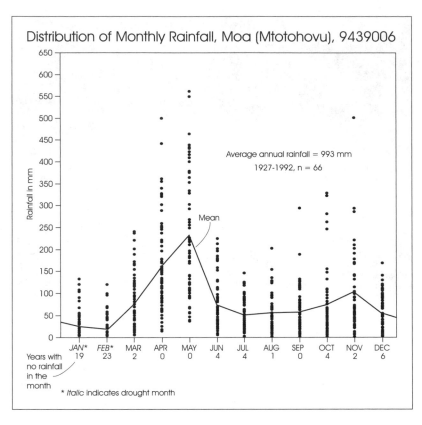

Distribution of Monthly Rainfall, Moa (Mtotohovu), 9439006

Average annual rainfall = 993 mm
1927-1992, n = 66

Mean

Years with
no rainfall
in the
month
* *Italic* indicates drought month

	JAN*	FEB*	MAR	APR	MAY	JUN	JUL	AUG	SEP	OCT	NOV	DEC
	19	23	2	0	0	4	4	1	0	4	2	6

FIG. 5.35. Rainfall diagram, Moa. (Source: Data from Tanzanian Meteorological Department)

at the time of sale of their catch because middlemen take a large cut. If a kilogram of fish was sold for 5 shillings in town, the fisherman, after paying rental fees, realized only about 1 shilling (Freyhold 1972: 2).

Rainfall in Moa (fig. 5.35) averages 993 mm/year and is notably bimodal, with one peak in May and another in November. Although rainfall is less between May and November, it averages just over 50 mm/month. January and February are drought months.

The student researchers who did interviews in Moa in 1972 had consistent views about various aspects of Moa as a community. Moa was average in most respects, except for its notably poor soils and low potential for agriculture. Because it is really a fishing village, it is probably the most distinctive village among the 18 we studied. There was only moderate hardship or risk in livelihood, in part because fishing can be done every

TABLE 5.18. Typical Planting Dates and Growth Periods, Moa

Rainy season	Crop	Average planting date	Days to maturity	n^a
Masika	Cassava	15 Mar.	350	10
	Coconuts	10 Mar.	6 yr	1
	Cowpeas	18 May (varies)	70	2
	Green grams	27 Apr.	72	5/4
	Maize	25 Mar.	96	2
	Rice	3 Apr.	116	2
	Sorghum	6 Apr.	78	1
	Sweet potatoes	12 Apr. (varies)	220	2
Mchoo	Cassava	6 July	350	2
	Cowpeas	5 July	—	1
	Maize	25 July	111	1
Vuli	Cassava	varies	350	2
	Cowpeas	1 Nov.	82	2
	Green grams	1 Nov.	75	1
	Maize	1 Nov.	105	1

Note: Characteristic mixes or intercroppings: *masika* rains—cassava and green grams; *mchoo* rains—cassava, green grams, and maize; *vuli* rains—maize, cowpeas, and cassava.
[a] 5/4 means planting date was based on 5 observations and season length was based on 4.

day throughout the year and the catch sold at the market. Given the poor soils, the potential for agriculture lies with tree crops such as cashews and coconuts. In the period 1972–1993 there was very little change in crops grown in Moa. By 1993 greater emphasis was being placed on cashews and coconuts, and a bit less on cowpeas and green grams.

The status of women in Moa is notably low. We noted in 1993 that most of those interviewed had been divorced, some more than once. Overall, one can say that Moa is a distinctive village, and its economic character reflects a reliance on fishing and perennial tree crops rather than annual crops.

The houses in Moa line both sides of a broad main street that runs perpendicular to the coast, almost to the shore. Other houses lie behind this main street, approximating a gridded street pattern. There has been much growth in the past two decades, and formerly agricultural land has been built upon. Some of the houses in the older part of the village are quite old and have carved doors reminiscent of the Arab/Indian architecture of Bagamoyo and Zanzibar Town.

At first I thought various kinds of modernization had occurred in Moa over two decades—television, electricity, phone service, fish refrigeration, piped water—but on closer inquiry I found out as follows. During the time of *ujamaa* (which started in June 1971), the villagers got a community fishing boat and new nets with the help of the government. For a time refrigeration for the fish catch was provided for the Moa Ujamaa Village Cooperative, but when the *ujamaa* village failed and was unable to pay its debts, the government came in and removed all the cold-storage equipment. Perhaps the project was doomed from the start, because the *ujamaa* approach had several flaws. First, it was not able to accommodate all the fishermen that wanted to participate but was limited to 35 registered members; and second, the technical services and equipment (cold storage, motorboats, etc.) were provided in a "top-down" manner. Government staff managed administrative and technical matters and did not train the local people adequately (Freyhold 1972: 4, 6).

The "television aerial" I saw turned out to be a solar-powered microwave transmitter that connects a public telephone at the Moa village office headquarters with the rest of Tanzania. While we were there, the phone rang several times (but no one bothered to answer) and several calls were made, one entirely in English. There is no electricity in the town, although there is electricity along the line of road leading from Tanga to Mpakani-Horohoro (on the Kenya border), probably for security considerations and on behalf of all the sisal estates along the road. The water supply in Moa is from boreholes. I saw one spigot and standpipe on a round concrete base, but it was not working. Others apparently were.

Summary and Analysis

This introduction to the 18 villages distributed across Tanga Region, from well-watered Kwadundwa in the west, with its single peak of rainfall (April), to Moa on the coast, with its double peak of rainfall (May and November), shows that the villages are quite varied in terms of resource endowment, settlement form, degree of isolation, involvement in commercial agriculture and other enterprises, and village dynamism and esprit. The villages can be placed in four groups: (1) those that appeared to have made substantial advances over the period 1972–1993, (2) those that had made some progress during the period, (3) those that were pretty much unchanged, and (4) those that seemed worse off.

Among those villages that appeared to have made remarkable progress, I would place Mzundu, Vugiri, Kisiwani, Magoma, and Daluni. Vugiri, Kisiwani, and Daluni have good access and excellent resource endowment; Mzundu has a good natural setting and profoundly improved accessibility, with the Chalinze-Segera tarmac highway now only 13 km away; Magoma also has good access and has been the site of much governmental improvement.

Those villages that had showed some improvement were Mgera, Kwamsisi, Mkomazi, Kisiwani, Maranzara, and Moa. All of these villages (except Kwamsisi) are connected by all-weather roads to the rest of Tanga Region, and they experienced urban growth in the period 1972–1993. Kwamsisi's improvement cannot be explained by resource endowment and accessibility, because these were little changed since 1972. My personal opinion is that the fruits of *ujamaa* (infrastructural investments) really did benefit Kwamsisi and that this isolated village had skillful leadership and a sense of esprit and self-reliance.

Those that seemed relatively unchanged since 1972 were Mandera, Mlembule, Kiwanda, and Mwakijembe. Mandera, surrounded by sisal estates, had little land for expansion. The collapse of the sisal industry meant that fewer jobs were available, as well as fewer opportunities to sell surplus agricultural produce to workers. The proximity of Korogwe Town, with employment opportunities, suggests that Mandera may be a "bedroom suburb," with little sense of its own cohesion as a community. Mlembule had a bad experience with *ujamaa,* and with competition from nearby Jitengeni and Mombo, there was little need to develop its own shops and services. In Kiwanda *ujamaa* was not a success. People were very suspicious of outsiders. The regional training college did not serve local people. Mwakijembe remains climatically risky. Keeping livestock was still the primary activity. Road connections to the Daluni-Maramba-Gombero-Tanga trunk road were still quite poor and there was no regular bus service.

Those that seemed worse off were Kwediamba, Minazini, and Kwamgwe. Kwadundwa could not be judged, since we did not visit it in 1993. The same features—resource endowment and transport access—can be invoked in explaining, as well, those villages that seemed worse off. Kwediamba and Minazini suffered from their very poor resource base and, paradoxically, their proximity to Handeni and Chanika. One could argue that they are not sufficiently isolated to have developed a sense of self or community. The commercial facilities of Chanika and the governmental facilities of Handeni Town are within easy walking distance. Kwamgwe is surrounded by idled sisal estate land. It was thus not an outlet for employment or sale of surplus produce to sisal employees. A number of recent immigrants are retired government servants, who are not part of the community.

On the whole, perhaps this account of which villages improved, stagnated, or declined between 1972 and 1993 is not a bad record, but the
amount of progress in villages in general was small in relation to the profound levels of poverty that still persist throughout the region.

Although I save the analysis of agricultural change over a 40-year span
(1961–1993) for chapter 7, I will list by way of summary a few salient points
here. The 182 farmers interviewed took up and abandoned (or grew more
or less of) a great many crops. On average each farmer grew more of at
least one crop and started growing one or two new crops. For the most
part, these crops were maize and cassava or tree crops (coconuts, cashews,
oranges, tangerines, jackfruit, bananas, or mangoes). Also, some farmers
grew more pineapples and cardamom. The villages whose farmers gave
greater emphasis to tree crops were either well endowed or well located
to move crops to market. Southwest to northeast these are Kwediamba,
Kwamsisi, Mzundu, Kwamgwe, Maranzara, Daluni, and Moa.

It is easier to say which villages did not show great increases in maize and
cassava cultivation. For maize, these were Mlembule, Mkomazi, and Moa.
For cassava, these were Mgera, Kwediamba, Mzundu, Mandera, Mlembule, Magoma, Mwakijembe, and Moa.

As to crops that have been abandoned or planted less, in general it has
been legumes and vegetables that have declined almost everywhere. This
is a worrisome trend, because some very nutritious crops are involved.
These include legumes (beans, cowpeas, green grams, pigeon peas, and
peanuts), oilseeds (such as simsim, castor, and cashews), and garden vegetables (onions, tomatoes, sour tomatoes, okra, spinach, pumpkins, and
sweet potatoes).

In a state of nature, Tanga Region would be made up of different kinds
of forests, ranging from montane cloud forests in the mountainous areas, to
deciduous forests at middle and lower elevations, to *Acacia* and *Brachystegia* woodlands and thickets in the drier areas (Hamilton 1989; National
Research Council 1982: 145; Trapnell and Langdale-Brown 1969). The
most ecologically sound crops for such areas are tree (or treelike) crops:
coconut, citrus, mango, pawpaw, jackfruit, and cashew trees in the lowlands
and bananas, coffee, tea, avocado, cardamom, and teak in the highlands.
The fact that more farmers are giving greater attention to tree crops that
can be sold is an encouraging sign.

Progress or stagnation in these 18 villages has taken place contempo-
raneously with Tanga Region's considerable population growth. In 1967
Tanga's population was 771,090; in 1988 it was 1,280,212 (an increase of

66 percent). By 2002 the population had increased to 1,642,015, a 28 percent increase. The greatest recent growth was in Handeni District, which increased its population from 1988 to 2002 by 57 percent. In some places, larger numbers of people in a village have led to increased economic activity and diversification (Daluni and Magoma). In other places, added people may have simply increased the local burden of poverty (Minazini and Kwediamba).

Assessing Different Farm Management Practices

I began this book by noting that there were five kinds of decisions or actions farmers could take to meet crop water needs. These actions can be objectified (and thereby modeled) as a sequence of choices: (1) place to grow crop, (2) which crop(s) to grow, (3) date to plant, (4) spacing of plant material (seeds or cuttings) and subsequent thinning, and (5) provision of supplemental water, through irrigation, water harvesting, etc. This chapter explores each of these topics using simulations provided by the model described in chapter 1 and Appendix A. I am asking five specific questions about water in relation to peasant farming in Tanga Region. It is important to understand the answer to each question if one wants to know why farmers succeed or fail.

Introducing the Energy–Water Balance Model

There are essentially three sorts of agroclimate, or energy-water balance, models (Baier 1979; Corbett 1990: 25). *Empirical models*, usually based on multiple regressions, generally have simple data requirements and are inexpensive to run. The results are, however, difficult to interpret. One cannot pinpoint the biophysical causes or events that resulted in different yields. *Parametric simulation models* are grounded in the biophysical processes occurring in the atmosphere, soil, and plant tissues. Their data requirements are much greater, as are their requirements for computer and programming capabilities. The causes of variations in simulated yields can more readily be observed. A third approach is the *economic model*, which seeks to determine the trade-offs for farmers in adopting particular practices, such as increasing the amount of nitrogen fertilizer in a field. The costs of application are offset by the benefits in the cash value of the yield obtained. The member institutions of the Consultative Group on International Agricultural Research (CGIAR) have regularly used economic models in their research and extension (CIMMYT 1988; International Institute of Tropical Agriculture 1999). A parametric simulation model is used in this study, and I ask the reader's indulgence in introducing it in the following way.

In Thomas Mann's *The Magic Mountain*, the central character, Hans Castorp, is referred to by the author (the novel's kindly, omniscient narrative guide) as "Life's delicate child." "Life's delicate child" is a useful metaphor for this book about Tanga's agriculture. The plant growing in the field is "Life's delicate child" and it is the farmer's ongoing concern for its well-being and success that forms a major focus of livelihood. Consider this tender plant. Who is it? Maize, sorghum, green grams? Where is it grown? In what soil, at what slope? What of its neighbors? How is it spaced among other plants? And what of events in its life cycle? Was it planted at a good time? Did it experience problems or damage in its youth because of inadequate or excessive rain, agricultural vermin (bushpigs, etc.), or disease? Was there stress at its time of flowering? (There is evidence in Mann's novel that Hans did indeed sleep with Clavdia Chauchat.) Did the plant mature and ripen successfully or wither and die in a great drought? (That question is inconclusive in *The Magic Mountain*, for when last seen, Hans is staggering out of sight in the smoke, tumult, and confusion of a World War I battlefield.)

The events surrounding a plant's life can be measured and modeled. We can tell the life story of individual plants and of entire communities (fields) of plants. In fact, the model used in this study keeps a careful diary and notes a host of events and conditions day by day in the life of the plant, and implicitly in generations of plants as the seed from one season is planted out in a subsequent season. (We have to move our metaphor to *Buddenbrooks* or John Galsworthy's *Forsythe Saga*.) There is still another novelistic feature of this study, the fictional counterfactual "what if?" We stand with reference to Tanga's agriculture as the novelist does with reference to the story being told. We can write different versions. (There is or is not a terrible blizzard. Frau Chauchat does or does not lend Hans her silver pencil.) In the agrometeorological simulations that follow, we rerun history, using different assumptions. This enables us to see what different outcomes ride with the choices farmers make.

In 1961–1962, I spent 13 months in East Africa—Kenya, Tanganyika (as it then was), and Uganda—as part of a research team, composed of anthropologists, in a project called Culture and Ecology in East Africa. My task was "to examine the potentialities of the environment for human exploitation and the manner in which such exploitation was accomplished" (Goldschmidt 1976: 6).[1] It soon became obvious that the most important factor affecting the dominant livelihood, farming, was rainfall—in its amount, its timing, and its variability. The same could be said about rain with respect to the second most important livelihood, the keeping of livestock.

I have spent several decades trying to understand and "operationalize" the interrelations of "environment" and "human exploitation" of it. These interrelations consist of flows of energy and moisture in the biosphere—the atmosphere, soils, and plants; and they involve the farmer as one who intervenes on behalf of plants.

The interactions of energy and water in the biosphere are known as the energy-water balance or budget. One studies the fluxes of energy and moisture in the atmosphere, soil, and plant tissues. Plants experience moisture stress at times when needed crop water is not available. The result is crop failure or a reduced yield (fig. 6.1). This too is studied as part of energy-water budgets. My focus in this study is on the actions farmers take on behalf of the crops they grow that are related to moisture demand and supply. There are, of course, many other actions a farmer takes to protect and nurture the crop, such as terracing, weeding, using fertilizers, and protecting against insects, plant pathogens, and agricultural vermin. The harvest and postharvest storage are other aspects of a farmer's management of the crop. These latter considerations are not the subject of this book. Here we explore farmer decisions and interventions that are directly related to crop water needs.

Early models of this nexus of energy-water-crop management used monthly rainfall and energy values. The research was aided and informed by the writings of Thornthwaite (1948), Manning (1956), McCulloch (1965), Penman (1948, 1963), Wang'ati (1968), Monteith (1965, 1972), Dagg (1965), Dupriez (1964), and Cochemé and Franquin (1967). Thus, if a maize crop grew over a 5-month season, one examined the crop-water

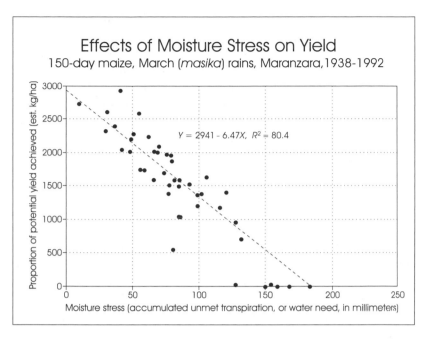

FIG. 6.1. Effects of moisture stress on yields of 150-day maize, Maranzara.

need (transpiration) and crop-water supply each month. If crop-water need exceeded crop-water supply, a deficit affecting the yield was the presumed result. The main drawback to monthly values was that it gave farmers (or the model mimicking farmer behavior) 12 days out of 365 on which to plant—1 January, 1 February, etc. This clearly is unrealistic.

To improve the model I switched to daily data. In 1978–1979, I spent a year in the Physics Division of the Kenya Agricultural Research Institute, where I worked closely with J. Ian Stewart (1980, 1982, 1988; Stewart et al. 1973; Stewart and Hash 1982) and Fred Wang'ati (1968, 1969, 1972), and also in Katumani and Kimutwa in Machakos District. Going to daily data expanded data requirements 30 times, but computers were also expanding capabilities. The model was calibrated through experimental trials of maize and sorghum (Darrah and Penny 1974; Darrah 1976; Eberhart, Penny, and Harrison 1973). I also incorporated new methods of tracking soil moisture (Rawls 1982; Ratcliff, Ritchie, and Cassel 1983; Gupta and Larson 1982). The model status in 1983 is given in Porter 1983.

The model we are discussing here bears similarities to numerous other agroscience or crop-water simulation models that have been developed to track the flows of energy and moisture in the atmosphere, soil, and plants and to estimate yields. Such models developed greatly in the late 1960s and early 1970s (Acock 1989). CERES, a model developed for maize, was one of the earlier crop-water models (Jones and Kiniry 1986; Thornton et al. 1995). Much of this work was coordinated by the International Benchmark Sites Network for Agrotechnology Transfer (IBSNAT) (Uchara et al. 1993). This organization was succeeded by the International Consortium for Agricultural Systems Application (ICASA) and work by the individual research institutes of the CGIAR (ISNAR 1996). Models for a wide variety of crops have been created. Among more recent ones are SWAP (Soil Water Atmosphere Plant), SWATRE (Soil Water Actual Transpiration Rate Extended), and SWACROP (Soil Water and Crop Production Model), developed by scientists at Wageningen Agricultural University in the Netherlands (Belmans, Wesseling, and Feddes 1983; Kabat, van den Broek, and Feddes 1992; Pereira et al. 1995). Other models deal with soil water or plant nutrition (Pilbeam, Daamen, and Simmonds 1995; Molina and Smith 1997). Increasingly remote sensing and GIS (geographic information systems) are being used in crop modeling (Diak et al. 1998; Kustas and Norman 1999; Michaud et al. 1999). The model used in this study is purposely limited to analysis of the consequences of a plant's receiving or not receiving needed moisture.

Since 1983 I have used a model based on the Food and Agriculture Organization approaches of Doorenbos and Kassam (1979) and Doorenbos and Pruitt (1977). The daily model, written in FORTRAN, was refined and improved (rewritten in BASIC) in the late 1980s by John D. Corbett (Corbett 1990). The program simulates crop growth and estimated yield, season by season, keeping track, day by day, of 12 interrelated variables (rainfall; fallow, or non-crop-period, runoff; soil saturation runoff; infiltration; transpiration; direct evaporation; daily evaporation; crop water need; daily water deficit; daily yield reduction; moisture storage; and water balance—the balance between all moisture incomes and outgoes). Details on the program are provided in Appendix A. The program produces a table at the end of its simulation summarizing a number of variables (table 6.1), and it permits one to show graphically (1) the overall results of the simulation, that is, the estimated crop yields (fig. 6.2), and (2) the daily interplay of rainfall, crop transpiration, soil moisture storage, yield reduction, and other variables for particular years. Figure 6.3 displays the results of simulations for two seasons at Minazini, near Handeni Town. I have chosen 1965 and 1966 because they illustrate two dramatically contrasted

TABLE 6.1. Summary of a Crop Simulation, Minazini, Handeni District

The percent of cropping season by stage: #.## .45 .1 .35 .1
The corresponding YRR for each stage: .4 1.1 .6 .35
CN fallow, early, late season 86 81 78
Station 9538007.pre 677 m; evaporation zone mombo.evp; AWC = 102
Season length = 120; crop = km120.crp; total E_o = 1546.104
Maximum yield = 4,560; soil suction restriction = .2; lag-0 days
Plant day window opens day 60; runoff store = 0 mm

Plt. year	Plt. day	Harv. year	Water need	Season precip.	Crop trans.	Season runoff	% trans.	% runoff	Est. yield	% yield
1928	72	1928	248	466	203	125	44	27	3,159	69
1929	85	1929	251	203	166	4	82	2	1,367	30
1930	72	1930	248	438	187	262	43	60	2,421	53
1931	77	1931	248	510	244	206	48	40	4,484	98
1932	74	1932	248	441	205	185	47	42	2,889	63
1933	80	1933	248	205	157	23	76	11	1,858	41
1934	73	1934	248	421	237	82	56	19	4,489	98
1935	72	1935	248	480	246	228	51	47	4,560	100
1936	94	1936	259	474	213	234	45	49	2,884	63
1937	72	1937	248	322	209	50	65	16	2,964	65
1938	81	1938	248	493	198	254	40	52	2,644	58
1939	72	1939	248	543	246	255	45	47	4,541	100
1940	75	1940	248	653	232	377	36	58	3,662	80
1941	77	1941	248	323	197	72	61	22	2,664	58
1942	72	1942	248	922	230	658	25	71	3,561	78
1943	96	1943	261	181	172	13	95	7	2,146	47
1944	72	1944	248	464	212	246	46	53	3,238	71
1945	111	1945	277	170	119	17	70	10	0	0
1946	121	1946	290	255	103	100	40	39	0	0
1947	72	1947	248	558	246	315	44	56	4,560	100
1948	88	1948	253	492	220	184	45	37	3,185	70
1949	94	1949	259	315	212	78	67	25	3,096	68
1950	72	1950	248	472	202	219	43	46	2,965	65
1951	72	1951	248	433	230	187	53	43	4,001	88
1952	90	1952	255	175	144	72	82	41	1,487	33
1953	75	1953	248	333	214	54	64	16	3,051	67
1954	100	1954	265	298	169	68	57	23	2,047	45
1955	83	1955	250	374	230	93	61	25	3,736	82
1956	72	1956	248	242	218	27	90	11	3,132	69
1957	72	1957	248	388	224	192	58	49	3,686	81
1958	72	1958	248	340	225	123	66	36	3,328	73
1959	72	1959	248	417	184	209	44	50	2,831	62
1960	72	1960	248	467	226	203	48	43	3,396	74
1961	72	1961	248	178	198	0	112	0	3,260	71
1962	72	1962	248	172	198	19	115	11	2,977	65
1963	72	1963	248	314	156	182	50	58	1,882	41
1964	72	1964	248	231	211	51	91	22	2,969	65
1965	110	1965	276	70	47	3	67	4	0	0
1966	72	1966	248	501	246	168	49	34	4,560	100
1967	72	1967	248	480	235	206	49	43	4,477	98
1968	72	1968	248	594	246	323	41	54	4,560	100
1969	92	1969	257	292	172	122	59	42	2,081	46

(*Continued*)

TABLE 6.1. (*Continued*)

Plt. year	Plt. day	Harv. year	Water need	Season precip.	Crop trans.	Season runoff	% trans.	% runoff	Est. yield	% yield
1970	72	1970	248	858	216	626	25	73	3,056	67
1971	82	1971	249	428	187	171	44	40	2,501	55
1972	72	1972	248	362	223	138	62	38	3,283	72
1973	85	1973	251	363	162	212	45	58	1,930	42
1974	95	1974	260	292	224	9	77	3	3,162	69
1975	72	1975	248	432	227	144	53	33	3,432	75
1976	72	1976	248	292	209	55	72	19	2,904	64
1977	72	1977	248	152	127	2	83	1	1,210	27
1978	72	1978	248	518	238	262	46	51	4,031	88
1979	77	1979	248	431	238	187	55	43	4,226	93
1980	103	1980	268	314	130	133	41	42	886	19
1981	75	1981	248	550	223	229	41	42	3,421	75
1982	74	1982	248	435	245	69	56	16	4,552	100
1983	85	1983	251	317	212	86	67	27	2,953	65
1984	91	1984	256	312	193	105	62	34	2,586	57
1985	91	1985	256	331	161	163	49	49	1,962	43
1986	88	1986	253	448	194	240	43	54	2,514	55
1987	72	1987	248	351	211	126	60	36	2,942	65
1988	76	1988	248	421	182	160	43	38	2,402	53
1989	73	1989	248	329	225	77	68	23	3,528	77
1990	72	1990	248	255	170	99	67	39	2,131	47
1991	85	1991	251	360	198	93	55	26	2,689	59
1992	94	1992	259	445	194	222	44	50	2,551	56

Note: See Appendix A for fuller explanation of the various parameters and assumptions for the simulation. In the above, line 1 refers to proportion of season that is vegetative, flowering, grain-filling, and ripening; line 2, YRR is yield reduction ratio; line 3, CN refers to US Soil Conservation Service curve number specifications, that is, the assumptions about soil moisture as the simulation begins; line 4, AWC is available water capacity, the product of soil depth times maximum soil water/meter; and line 6, soil suction refers to permanent wilting point.

seasons: 1965, when the crop failed totally, and 1966, when the rains were plentiful and 100 percent of potential yield was (likely) achieved. In 1965 the rains were late and inadequate. The search for a suitable planting date commenced on 1 March, but planting occurred only on 21 April, and the crop transpired only 47 mm during its entire season. There was a serious moisture shortage during the flowering period and through the remainder of the season. The following year, by contrast, planting occurred on 12 March, and 246 mm of the 248 mm needed for transpiration were supplied by ample rains.

The model permits us to return to the task cited at the beginning of this chapter: "to examine the potentialities of the environment for human exploitation and the manner in which such exploitation was accomplished." Admittedly, my approach to the task is narrowed to a consideration of energy-water flows in crops and farmer management of crops. In the

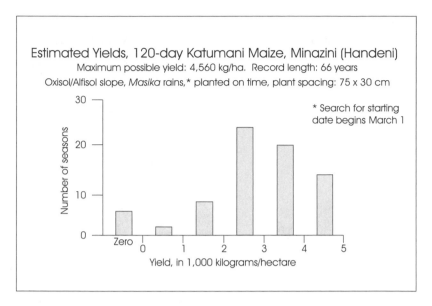

FIG. 6.2. Estimated yields of 120-day Katumani maize, Minazini, Handeni District.

model, I hold constant such aspects as soil fertility, soil nutrition, and the control of crop predators and pathogens (Yudelman, Ratta, and Nygaard 1998).[2] But even restricting ourselves to the five farmer-controlled variables cited above opens up a very large number of possible combinations in our simulations.

In order to assess the importance of a given variable (a farming practice, such as choice of crop, planting population, or delay in planting, or an environmentally given condition, such as moisture-holding characteristics of a particular soil), it is necessary to hold constant all other things that might vary. Thus, in the following analysis I hold constant all aspects of environment and management except the variable being studied. For example, I establish a standard soil moisture, planting population, and crop when I am studying delay in date of planting.

The complexities of crop cultivation in Tanga Region are great, and the simulation of the possible combinations of decisions a farmer could make is astronomical. This study is already complex. No reader would have the patience to follow the "changes" that could be "rung out" on 26 crop types or crop combinations, including differences in days to maturity, five soil types, 17 starting dates[3] over two seasons (*vuli*, October peak; and *masika*, April peak), four planting lag times (from no delay to 42 days late

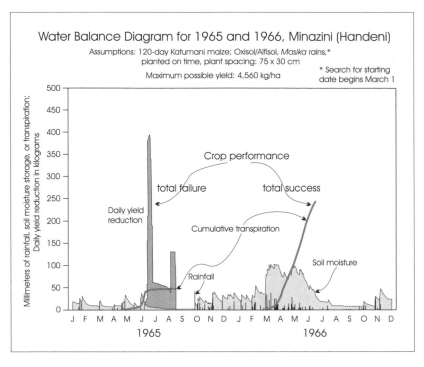

FIG. 6.3. Water-balance diagram for 1965 and 1966, Minazini, Handeni District.

in planting), and 11 plant populations. This list would eventuate in 97,240 studies of particular combinations in a place. Applied to all 18 villages in the transect, the number of potential, separate studies becomes 1,750,320, each with an outcome as to the estimated yield and the estimated reliability of that yield. This would be the ultimate "shaggy-dog story."

To gain some control over these materials, I have chosen to illustrate with a limited number of simulations the consequences of particular farmer choices, such as delay in planting, changing crop spacing, or growing a particular crop. My purpose is to demonstrate two things: (1) that a biophysically reasoned model of energy and moisture fluxes in atmosphere, soil, and plant gives one great flexibility in evaluating farmer strategies in particular places, and (2) that the way farmers manage their livelihoods reflects their extensive understanding of the requirements of their crops. Farmers, of course, are affected by other aspects of life, particularly Tanzania's political economy, and we have paid attention to that aspect as well (chapter 4). It should also be emphasized that the simulations are

TABLE 6.2. Available Water Capacity for Tanga Soils

| | | Range (mm/m) | |
Soil name	Mean (mm/m)	Upper	Lower
Oxisols/Alfisols upland ($n = 40$)	104	146	72
Oxisols/Alfisols slope ($n = 7$)	113	153	91
Oxisols/Alfisols base of slope ($n = 21$)	111	148	90
Vertisols ($n = 14$)	114	165	68
Entisols ($n = 4$)	110	138	77

only estimates of yields, based on the degree to which moisture needs were met. Yields are also affected by many other things, such as the application of manure and the depredations of pathogens, insects, and vermin.

Choice of Site and Soil

Originally, I thought that I would present results of simulations showing the differences in yield that result from planting on uplands, hillslopes, and valley bottoms, and in relation to different kinds of soil. The water-holding capacity of the soils tested (and listed in table 3.3) varied enormously but exhibited the means and ranges given in table 6.2. The average water-holding capacity of the soils is surprisingly similar among types, and there is little to be gained from an analysis of soils in different places along the catena—upland, slope, and base. The moisture-holding capacity of soils is important, however; and the difference between a soil that can hold 170 mm/m and a soil that can hold only 70 mm/m is worth examining in relation to performance of crops. One demonstration of the consequences of using a soil that holds much moisture versus a soil with limited capacity follows.

We simulated the growing of a crop of maize at Minazini (Handeni District), using contrasting assumptions about soil moisture-holding capacity. We grew the crop during the *masika* rains (search for planting date commenced on 1 March) using a hybrid maize crop that matures in 150 days. We kept every aspect of the simulation the same except for maximum soil moisture, 70 mm and 170 mm. Since the soil was assumed to be 1.2 m in depth, the maximum water-holding capacities in the two instances were 84 mm and 204 mm, a difference of 120 mm. The extra 120 mm (4.7 inches) makes a difference in a lot of ways (figs. 6.4 and 6.5).

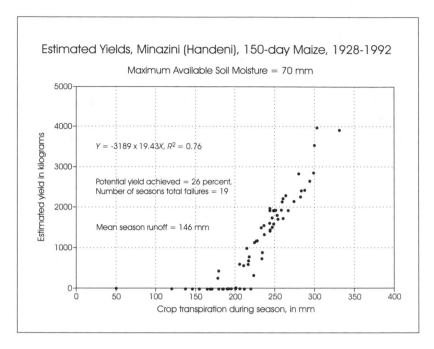

FIG. 6.4. Estimated yields of 150-day maize (70 mm), Minazini, 1928–1992.

Estimated average yield is 2,677 kg/ha versus 1,210 kg/ha (over twice as much); 58 percent of the yield potential is realized versus 26 percent. Of the 65 years of record, only 8 resulted in total crop failure with the higher soil moisture value versus 19 total crop failures using the lower value. The difference is also reflected in seasonal runoff, that is, water that could not be stored by the soil but ran off during storms or drained away. The simulations showed an average loss of only 90 mm for the soil with the higher water capacity versus 146 mm for the soil with lower water capacity. Farmers know which particular fields or parts of fields hold moisture well, and they choose them over soils with lower water capacities when they can.

Choice of Crop

Crops vary as to how many days they need to mature and how much water they need at each stage of their life cycle. The transpiration coefficients

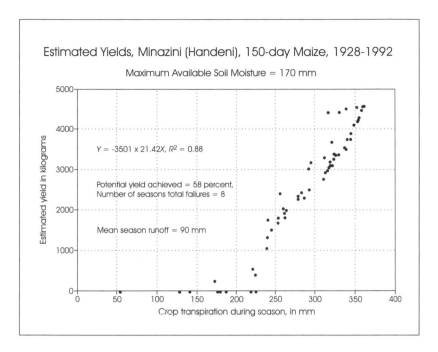

FIG. 6.5. Estimated yields of 150-day maize (170 mm), Minazini.

(E_t/E_o) of crops grown in East Africa have been determined for many crops by lysimeter and other field experiments (Blackie and Bjorking 1968; Brown 1963; Carr 1968; Dagg and Othieno 1968; Dupriez 1964; Nieuvolt 1973; Wang'ati 1968, 1972; Willatt 1968; D. A. Rijks, pers. comm., August 1969, Namulonge, Uganda). We can simulate growth of many crops and combinations or interplanting of crops. My analysis centers on maize because it is the most important crop grown by Tanga's farmers and because more research has been devoted to it than any other staple crop.[4] Some attention is given to interplanting and to crops besides maize.

If farmers plant 150-day hybrid maize under uniform conditions of soil moisture-holding capacity and soil depth (85 mm/m and 1.2 m in these simulations), the chance of realizing all of the potential yield ranges between 97.2 (Vugiri) and 1.7 (Mkomazi) in the *masika* (March) rains (table 6.3).[5] Besides Vugiri, five other places in central and eastern Tanga Region realize more than 50 percent of the potential yield (fig. 6.6). For the *vuli* (October) rains, the result is much poorer, ranging between 45.4 percent (Mgera) and 0.0 (Mkomazi), with most locations (12 of 16) obtaining less than 20 percent of the potential yield (table 6.4).[6] Efforts to

TABLE 6.3. Simulation for Hybrid Maize in Tanga, *Masika* (March) Rains

Code: E3K005 = 150-day hybrid maize, planted on time, soil depth 1.2 m, soil moisture 85 mm/m

Site name	Code	Est. av. yield (kg)	Times failed to begin season	No. total crop failures	Percent total crop failures	Percent potential yield obtained	r^{2*}	No. of years
Mgera	MG	2,082	0	5	18	38.0	0.84	28
Kwediamba	HA	1,738	0	9	14	38.1	0.83	65
Minazini	HA	1,738	0	9	14	38.1	0.83	65
Kwamsisi	KW	864	0	7	32	27.7	0.70	22
Mzundu	MZ	1,047	1	12	38	25.8	0.61	32
Kwamgwe	KM	2,029	0	3	10	57.3	0.95	31
Mandera	MS	1,819	0	7	5	46.9	0.81	19
Mlembule	MO	932	0	11	34	24.3	0.74	32
Mkomazi	BU	64	1	20	91	1.7	0.21	22
Vugiri	VU	5,878	0	0	0	97.2	0.84	63
Magoma	MA	1,296	0	5	29	34.5	0.78	17
Kisiwani	KI	1,989	0	0	0	63.7	0.87	35
Kiwanda	KI	1,989	0	0	0	63.7	0.87	35
Daluni	DA	1,902	0	3	15	51.4	0.92	20
Mwakijembe	MW	302	0	6	75	9.8	0.82	8
Maranzara	PG	1,485	0	5	11	49.5	0.81	45
Moa	MT	1,437	0	7	11	53.3	0.82	63
Mlingano	ML	2,477	0	1	2	64.6	0.92	58

Note: Data for Kwadundwa not available.
*Column of figures is r^2 of season crop transpiration and estimated percentage of maximum yield achieved.

grow hybrid maize in northeastern Tanga result in unmitigated disaster (e.g., Moa, Maranzara, and Mlingano).

Suppose farmers shift to a shorter-season, less water-demanding crop, such as 105-day Katumani maize. The quality of performance increases for most places, indeed, to over 60 percent of potential yields realized in the *masika* (March) rains for all places except Mkomazi and Mwakijembe (table 6.5). Ten places achieve 80 percent or more of potential yield. Indeed, that potential in kilograms/hectare is greater than the yield from hybrid maize. The average estimated yield of the 10 places is 2,820 kg/ha for the 105-day Katumani maize, as opposed to 2,284 kg/ha for the 150-day hybrid maize. Clearly, there are advantages to planting a crop that matures in three and a half months versus one that requires five months. Only in Vugiri, at an elevation of 1,219 meters, does the 150-day hybrid maize outperform the Katumani maize (5,878 vs. 4,840 kg/ha).

If farmers try to grow Katumani maize during the *vuli* (October) rains, they are much less successful than they are during the *masika* rains

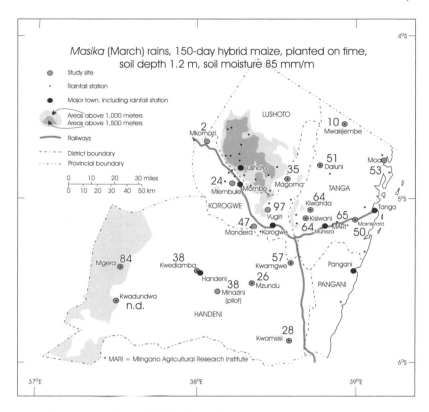

FIG. 6.6. Percentage of potential yield obtained.

(table 6.6). Only two places have good performances (Vugiri and Mgera). Using local practice as to variety and season lengths, we see that the performance for the *masika* (March) rains is better than that for the *vuli* (October) rains (tables 6.7 and 6.8). In most places, average yields are over 2,000 kg/ha and there are few total crop failures (table 6.7). Exceptions are Mkomazi, which depends on irrigation in any case, and Mwakijembe, which experiences frequent crop failures.[7] In the *vuli* rains (table 6.8), most places have average yields of 250–1,500 kg/ha, and the frequency of crop failure is quite high (except for Mgera, Mzundu, and Kisiwani). We can illustrate the relationship between season crop transpiration and estimated yield with two contrasted situations, where the percentage of potential achieved is high and where the percentage is low. Kwamgwe in *masika* and *vuli* rains shows this contrast nicely (tables 6.7 and 6.8 and fig. 6.7). In the *masika* rains, the average yield is 3,355 kg/ha, and there are no crop

TABLE 6.4. Simulation for Hybrid Maize in Tanga, *Vuli* (October) Rains

Code: E3V005 = 150-day hybrid maize, planted on time, soil depth 1.2 m, soil moisture 85 mm/m

Site name	Code	Est. av. yield (kg)	Times failed to begin season	No. total crop failures	Percent total crop failures	Percent potential yield obtained	r^{2*}	No. of years
Mgera	MG	310	1	2	10	45.4	0.63	21
Kwediamba	HA	891	5	25	42	21.5	0.53	59
Minazini	HA	891	5	25	42	21.5	0.53	59
Kwamsisi	KW	389	0	11	52	12.5	0.58	21
Mzundu	MZ	492	2	18	62	12.2	0.46	29
Kwamgwe	KM	474	0	15	50	13.4	0.64	30
Mandera	MS	518	3	10	67	13.3	0.39	15
Mlembule	MO	499	0	16	55	13.0	0.63	29
Mkomazi	BU	0	6	14	100	0.0	—	14
Vugiri	VU	2,329	2	12	20	38.5	0.77	59
Magoma	MA	86	2	11	79	4.9	0.25	14
Kisiwani	KI	567	0	15	71	18.2	0.65	31
Kiwanda	KI	567	0	15	71	18.2	0.65	31
Daluni	DA	748	2	10	56	20.2	0.69	18
Mwakijembe	MW	137	0	7	78	4.3	0.54	9
Maranzara	PG	207	0	30	71	6.9	0.38	42
Moa	MT	34	3	53	88	1.3	0.19	60
Mlingano	ML	391	2	33	60	10.2	0.54	55

Note: Data for Kwadundwa not available.
*Column of figures is r^2 of season crop transpiration and estimated percentage of maximum yield achieved.

failures ($r^2 = 0.92$); during the *vuli* rains, however, average yields drop to 1,126 kg/ha, and 46 percent of seasons result in crop failures ($r^2 = 0.90$). Note that the water demand for the *vuli* season is considerably greater (average 284 mm vs. 223 mm for the *masika* season).[8] The reason for this can be found by reference to figure 3.3, which shows the evaporation curve for Mlingano. The average starting day for crops during the *vuli* rains was 19 September. Some weeks later, when the crop had reached flowering stage, daily evaporation values averaged 6 or 7 mm/day. During the *masika* rains, the average planting date is around 9 March, and in the comparable period of greatest water need, daily evaporation values are only 4 or 4.5 mm/day.

Growing Enough Food

Farm families manage, on average, 1.45 ha of cropland each year. If they are to have sufficient food for their needs, they must harvest at least 180 kg of

TABLE 6.5. Simulation for Katumani Maize in Tanga, *Masika* (March) Rains

Code: B3K005 = 105-day Katumani maize, planted on time, soil depth 1.2 m, soil moisture 85 mm/m

Site name	Code	Est. av. yield (kg)	Times failed to begin season	No. total crop failures	Percent total crop failures	Percent potential yield obtained	r^{2*}	No. of years
Mgera	MG	3,805	0	0	0	86.9	0.94	28
Kwediamba	HA	2,953	0	3	5	80.9	0.88	66
Minazini	HA	2,953	0	3	5	80.9	0.88	66
Kwamsisi	KW	1,849	0	1	5	74.0	0.92	22
Mzundu	MZ	2,387	1	1	3	73.8	0.86	32
Kwamgwe	KM	2,462	0	0	0	89.2	0.98	31
Mandera	MS	2,630	0	1	5	84.8	0.94	19
Mlembule	MO	2,200	0	0	0	71.6	0.91	32
Mkomazi	BU	720	1	12	55	23.5	0.86	22
Vugiri	VU	4,840	0	0	0	100.0	1.00	63
Magoma	MA	2,085	0	2	12	61.6	0.97	17
Kisiwani	KI	2,314	0	0	0	92.7	0.92	35
Kiwanda	KI	2,314	0	0	0	92.7	0.92	35
Daluni	DA	2,548	0	0	0	86.0	0.97	20
Mwakijembe	MW	777	0	2	22	31.6	0.88	9
Maranzara	PG	2,038	0	0	0	84.9	0.94	45
Moa	MT	1,755	0	2	3	80.5	0.96	63
Mlingano	ML	2,850	0	0	0	92.8	0.96	58

Note: Data for Kwadundwa not available.

*Column of figures is r^2 of season crop transpiration and estimated percentage of maximum yield achieved.

grain per capita and obtain that yield reliably. Most grains, but particularly maize, do not store well in Tanga Region, and losses to insects and agricultural vermin, both during the growing season and in the postharvest period, are very high (Mascarenhas 1971a). The average family size in Tanga Region is 6.7, and all but three communities range between 5.9 and 7.8 persons per household. This means that their 1.45 hectares should produce, minimally, 1,206 kg of a staple food grain, or 831 kg/ha. In order to allow for differences in household size and provide some margin for food grain loss in storage, let us choose 1,500 kg per household, an average yield of 900 kg/ha.

If 900 kg/ha is the performance we are seeking from each place, some of the 17 communities fail even in the *masika* (March) rains, and even more communities fail during the *vuli* (October) rains (table 6.9). In the March rains, Mwakijembe and Mkomazi cannot be expected to achieve that minimal yield on average, and over 75 percent of years are likely to be a total loss. In the October rains, these two places are joined by Kwamsisi.

TABLE 6.6. Simulation for Katumani Maize in Tanga, *Vuli* (October) Rains

Code: B3V005 = 105-day Katumani maize, planted on time, soil depth 1.2 m, soil moisture 85 mm/m

Site name	Code	Est. av. yield (kg)	Times failed to begin season	No. total crop failures	Percent total crop failures	Percent potential yield obtained	r^{2*}	No. of years
Mgera	MG	2,623	1	3	14	60.0	0.58	21
Kwediamba	HA	1,336	0	14	24	36.7	0.64	59
Minazini	HA	1,336	0	14	24	36.7	0.64	59
Kwamsisi	KW	541	0	12	57	21.7	0.92	21
Mzundu	MZ	801	2	10	34	24.8	0.69	29
Kwamgwe	KM	752	0	7	23	27.3	0.67	30
Mandera	MS	616	3	7	47	19.9	0.64	15
Mlembule	MO	797	1	7	24	26.0	0.62	29
Mkomazi	BU	7	6	14	93	0.0	—	15
Vugiri	VU	3,001	2	3	5	62.1	0.77	59
Magoma	MA	291	1	10	71	9.6	0.56	14
Kisiwani	KI	921	0	7	23	37.0	0.92	31
Kiwanda	KI	921	0	7	23	37.0	0.92	31
Daluni	DA	1,025	2	7	37	34.6	0.82	19
Mwakijembe	MW	219	0	7	78	8.9	0.57	9
Maranzara	PG	359	0	23	55	15.0	0.76	42
Moa	MT	797	1	8	28	26.0	0.62	29
Mlingano	ML	990	2	18	33	32.2	0.82	55

Note: Data for Kwadundwa not available.

*Column of figures is r^2 of season crop transpiration and estimated percentage of maximum yield achieved.

Either an alternative kind of agriculture is essential (such as irrigation in Mkomazi), or an alternative or supplemental mode of livelihood is necessary (as with livestock keeping in Mwakijembe).[9]

Delay in Planting

Timing is an important aspect of farming. There is a large literature about the importance of planting on time in East Africa, that is, when the rains begin (Dowker 1963, 1964; Goldson 1963; D. J. Turner 1966; Akehurst and Sreedharan 1965). But just as important as planting on time may be the need to space out planting times so that limited quantities of labor can be applied in a way that makes farming possible. We explore this tension between these two demands of the agricultural system. The first example is a simulation of 150-day maize (*masika* rains) grown at Mlingano, which has a complete climatological record of 58 years (fig. 6.8). The distribution

TABLE 6.7. Simulation for Maize in Tanga, *Masika* (March) Rains

Code: Various maize varieties and season lengths based on local knowledge (statements in 1972 about season length and usual planting date for maize), but planted on time, soil depth 1.2 m, soil moisture 85 mm/m

Site name	Code	Est. av. yield (kg)	Times failed to begin season	No. total crop failures	Percent total crop failures	Percent potential yield obtained	r^{2}*	No. of years
Mgera	MG	4,614	0	0	0	84.4	0.57	28
Kwediamba	HA	no analysis						
Minazini	HA	no analysis						
Kwamsisi	KW	2,306	0	1	5	74.0	0.92	22
Mzundu	MZ	2,656	1	1	3	65.8	0.55	32
Kwamgwe	KM	3,355	0	0	0	94.8	0.92	31
Mandera	MS	2,898	0	1	5	74.8	0.94	19
Mlembule	MO	2,170	0	3	9	56.6	0.88	32
Mkomazi	BU	64	1	20	91	1.7	—	22
Vugiri	VU	6,052	0	0	0	100.0	1.00	63
Magoma	MA	1,353	0	3	18	36.1	0.88	17
Kisiwani	KI	2,652	0	0	0	85.0	0.91	35
Kiwanda	KI	2,652	0	0	0	85.0	0.91	35
Daluni	DA	2,776	0	0	0	74.9	0.85	21
Mwakijembe	MW	215	0	6	75	7.0	0.52	8
Maranzara	PG	1,485	0	5	11	49.5	0.81	45
Moa	MT	2,503	0	0	0	88.5	0.96	63
Mlingano		no analysis						

Note: Data for Kwadundwa not available.

*Column of figures is r^{2} of season crop transpiration and estimated percentage of maximum yield achieved.

of yields is given for planting on time and with lags or delays of 14, 28, and 42 days (2, 4, and 6 weeks). The number of failed seasons (the separate zero column) rises steadily from 5 to 23 seasons. Being late by 6 weeks is no worse than being late by 4 weeks. Indeed, there were a couple fluke seasons when yields greater than 3,000 kg/ha were achieved. A delay of 2 weeks at the beginning of the rains has a marked effect on yields, reducing estimated average yields from 1,613 kg/ha to 1,148 kg/ha and increasing the number of failed seasons from 5 to 13.

The next example is from Mgera. Ironically, although this village (among the 18 in the transect) is closest to having a unimodal rainfall, people distinguish two rainfall seasons, *vuli* and *masika* (the latter is also sometimes called *mwaka*). The two seasons overlap somewhat since December is a kind of transitional month, when the *masika* rains may or may not have started. The analysis of the cropping patterns shows interesting regularities regarding intercropping and associations of particular crops with *vuli*

TABLE 6.8. Simulation for Maize in Tanga, *Vuli* (October) Rains

Code: Various maize varieties and season lengths based on local knowledge (statements in 1972 about season length and usual planting date for maize), but planted on time, soil depth 1.2 m, soil moisture 85 mm/m

Site name	Code	Est. av. yield (kg)	Times failed to begin season	No. total crop failures	Percent total crop failures	Percent potential yield obtained	r^{2*}	No. of years
Mgera	MG	3,051	0	2	13	55.8	0.55	15
Kwediamba	HA	no analysis						
Minazini	HA	no analysis						
Kwamsisi	KW	1,112	0	7	33	35.7	0.93	21
Mzundu	MZ	382	0	2	7	9.4	0.52	30
Kwamgwe	KM	1,126	2	13	46	31.8	0.90	28
Mandera	MS	428	3	9	60	11.1	0.51	15
Mlembule	MO	no analysis (season not used)						
Mkomazi	BU	8	8	15	94	0.00	—	16
Vugiri	VU	3,094	4	14	24	51.1	0.73	58
Magoma	MA	263	3	12	80	7.0	0.46	15
Kisiwani	KI	1,514	0	3	10	48.5	0.89	31
Kiwanda	KI	1,116	0	8	26	35.8	0.90	31
Daluni	DA	1,277	3	6	33	34.5	0.77	18
Mwakijembe	MW	241	0	7	78	7.9	0.55	9
Maranzara	PG	301	2	29	71	10.1	0.60	41
Moa	MT	263	4	41	68	9.3	0.70	60
Mlingano	ML	no analysis						

Note: Data for Kwadundwa not available.

*Column of figures is r^2 of season crop transpiration and estimated percentage of maximum yield achieved.

and *masika* rains. Cassava is often interplanted with cowpeas and green grams (*choroko*). Maize and castor are commonly interplanted. Crops that are grown in relatively pure stands are bananas, beans, tobacco, and sweet potatoes. Even cassava, cowpeas, green grams, and maize are often grown in pure stands. Table 6.10 gives percentages of the crop reported as being grown in mixed stands by respondents in Mgera.

The second feature of the agricultural system is that cassava, cowpeas, and green grams dominate the planting schedule during the *vuli* rains, whereas maize and beans dominate the *masika* rains. During the *vuli* rains, green grams are planted out in late November, followed by cassava and cowpeas in early and mid-December, and then by some maize. About two-thirds of the interplanting takes place during the *vuli* rains. Of course, since cassava takes about a year to mature, it makes sense to plant another crop during the period when cassava is establishing itself. Both cowpeas and

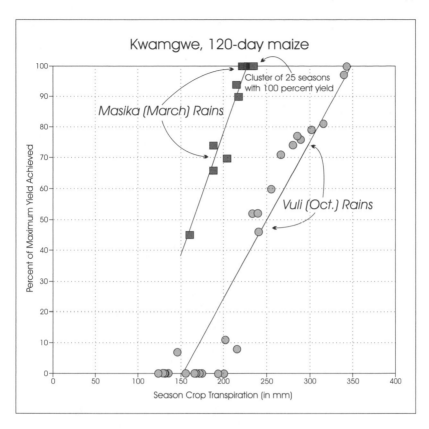

FIG. 6.7. Comparison of *masika* and *vuli* rains.

green grams take about 90 days to mature. The maize planted during the *vuli* rains takes about 125 days.

The *masika* rains feature maize and beans, planted separately for the most part or with beans planted in maize fields that were planted earlier. The maize is generally planted at the very end of January and takes just over 100 days to mature. Beans are planted around mid-April and take 85 days to mature. Sweet potatoes are planted in early to mid-May and are ready for use in about 125 days.

Many other crops are reported among the farmers of Mgera, but not with sufficient frequency to make it possible to establish usual planting dates. Among crops reported are oranges, sugarcane, Irish potatoes, onions, sorghum, and "vegetables."

TABLE 6.9. The Performance of Villages in Providing Adequate
Food

Site name	Code	Masika (March) rate of failure (percent)	Vuli (October) rate of failure (percent)
Mgera	MG	4	14
Kwediamba	HA	6	41
Minazini	HA	6	41
Kwamsisi	KW	9	71
Mzundu	MZ	9	66
Kwamgwe	KM	3	60
Mandera	MS	5	73
Mlembule	MO	9	59
Mkomazi	BU	82	100
Vugiri	VU	0	14
Magoma	MA	18	86
Kisiwani	KI	0	52
Kiwanda	KI	0	52
Daluni	DA	10	53
Mwakijembe	MW	67	89
Maranzara	PG	2	81
Moa	MT	21	92
Mlingano	ML	0	55

Note: Data for Kwadundwa not available.

The practice of recognizing two rainy seasons, whereas meteorologically it appears to be one, here at the *western* end of our transect, coupled with the variety of crops planted, means that farmers are active during many months. Although there may be some errors in the extreme values in planting dates reported to us (and the earliest and latest were ignored in establishing planting dates and season lengths), the range of dates does suggest that farmers are active over many months in their agriculture. Figure 6.9 shows the overlapping of planting dates reported for each mention of a crop, and thus the figure implicitly reveals labor inputs, which, as can be seen in the lower graph, are spread over a period from October (a time of clearing) to September (when the last crop is harvested). Many farmers made the point that there was great latitude in planting for the *vuli* rains. Although it was best to plant with the first good rains, all was not lost if one failed to do so. People were more relaxed about the *vuli* planting than they were about the *masika* planting; it had no rigid deadline. A planting might be done a little at a time, week by week, as the season advanced. The *vuli* rains and the period leading up to the *masika* rains are

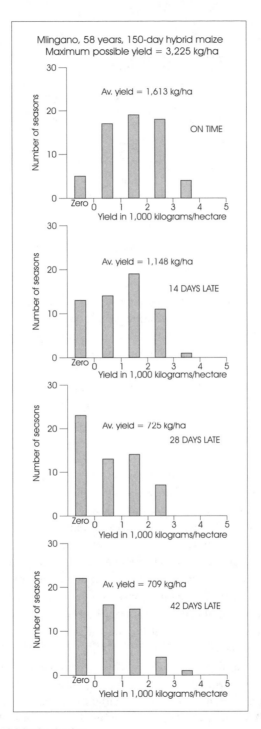

FIG. 6.8. Effects of delay in planting.

TABLE 6.10. Percentage of Crops Grown in
Mixed Stands, Mgera, Handeni District

	Mixed	Pure stands
Bananas	0	100
Beans	0	100
Cassava	67	33
Castor	50	50
Green grams	43	57
Maize	38	62
Sweet potatoes	0	100
Tobacco	0	100

so variable and unknowable that the practice of spreading out planting has a kind of crop insurance value.

Handeni Town provides a third example of the consequences of delay in planting (table 6.11). For this town, delay in planting during the *vuli* rains does not have serious consequences. In fact, delay may actually increase yields if it carries a crop over into the beginning of the *masika* rains. The simulation started its search for a planting date on 15 October, but the average planting date that resulted was 11 November. If one delayed the *commencement* of the search for a planting date until 30 November (42-day lag), the average planting date was 25 December and this delay increased crop performance, both for 90-day Katumani maize (9.2 percent to 12.8 percent of potential yield) and for 150-day hybrid maize (5.6 percent to 21.6 percent).

In the *masika* (March) rains, however, timing of planting is crucial. If one plants 90-day Katumani maize on time, the reward is 72.6 percent of estimated potential yields, whereas a delay of 6 weeks reduces potential yields to 13.4 percent. The consequences of delay in planting 150-day hybrid maize are much more serious, and maize that takes this long should probably not be grown. If the crop is planted on time, only 10.3 percent of potential yield is achieved; and if one delays 6 weeks, the yields, season after season, will be nil. The only consequence of delay in planting in the *vuli* rains may be that it can create a labor bottleneck. The farmer may be trying to plant a crop for the *masika* season at a time when *vuli*-planted crops need to be harvested.

The farmers in almost all villages chose their planting dates well for the *masika* rains. (Details on the timing of plants and season lengths of maize varieties chosen in the various villages for *masika* and *vuli* rains are given in Appendix C.) They could have done better in Mwakijembe

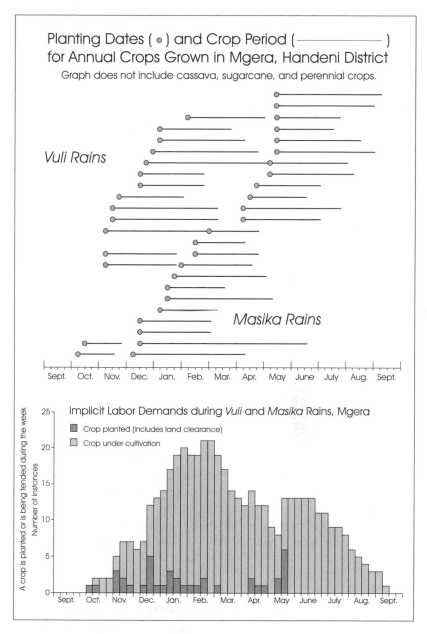

FIG. 6.9. Planting dates and implicit labor demands for crops grown in Mgera, Handeni District.

TABLE 6.11. Effects of Planting Delay in Handeni Town for Two Crops

Timing	Av. yield (kg/ha)	Percent realized	Av. day planting begins	Date search begins
Katumani (90-day) maize				
Vuli rains				
On time	334	9.2	11 Nov.	15 Oct.
14-day lag	278	7.7	23 Nov.	29 Oct.
28-day lag	345	9.5	10 Dec.	16 Nov.
42-day lag	464	12.8	25 Dec.	30 Nov.
Masika rains				
On time	2,648	72.6	21 Mar.	1 Mar.
14-day lag	1,981	53.2	4 Apr.	15 Mar.
28-day lag	1,128	30.9	18 Apr.	29 Mar.
42-day lag	488	13.4	2 May	12 Apr.
Hybrid (150-day) maize				
Vuli rains				
On time	254	5.6	11 Nov.	15 Oct.
14-day lag	407	8.9	23 Nov.	29 Oct.
28-day lag	718	15.7	10 Dec.	16 Nov.
42-day lag	986	21.6	25 Dec.	30 Nov.
Masika rains				
On time	465	10.3	21 Mar.	1 Mar.
14-day lag	98	2.2	4 Apr.	15 Mar.
28-day lag	12	0.3	18 Apr.	29 Mar.
42-day lag	4	0.1	2 May	12 Apr.

and Daluni by planting earlier. In four villages the farmers might have done better during the _masika_ rains by growing a crop with a different time to maturity—shorter in Maranzara, Mwakijembe, and Magoma, longer in Kwamgwe, since yields on a longer-growing crop are potentially higher.

The _vuli_ rains present a host of problems in timing and days to crop maturity. Many farmers solved the problem by not using the _vuli_ rains at all, or as much. On average, two more kinds of crops were planted in the _masika_ rains than in the _vuli_ rains, and for every farmer involved in planting a crop in the _vuli_ rains, nearly two farmers (1.7) were planting a crop in the _masika_ rains.

We have devised a crude measure of the degree to which farmers made good decisions or bad decisions on planting date and days to crop maturity in relation to the simulations of maize. The measure was made (using information presented in the crop calendars, chapter 5) by assigning a plus (+) to all farmers whose decisions agreed with those recommended by the simulation and a minus (−) to all farmers whose decisions disagreed with the simulation's recommendations. (It should be noted that date of planting and crop season length are averages of the data the farmers provided. Many farmers stressed that there was considerable flexibility in planting date, especially during the *vuli* rains.) The measure gave the following results: 84 percent of farmers chose appropriate planting dates in the *masika* rains, and 68 percent chose a maize variety with an optimal growing period. For the *vuli* rains, 74 percent of farmers chose appropriate planting dates, but only 52 percent chose a crop with an appropriate growing period. Although the measure's use of average planting date and average season length masks variation in actual practice, there appears to be room for improvement in choice of days to maturity for maize, as well as planting dates, for a number of farmers in our study. In the absence of simulations for other crops, one can only assume that the farmers make equally good or faulty decisions on the timing of planting and the season lengths of the crops they choose to grow.

There is an interesting geography of the timing and presence and/or absence of various pulses (beans, cowpeas, and green grams) in the 18 villages. Beans are very much a *masika* season crop, grown everywhere except in the extreme northeast (Mwakijembe, Moa, and Maranzara). Only a few farmers grow beans during the *vuli* rains, and these are mainly at higher elevations. Everywhere except Kisiwani, cowpeas are grown most prominently during the *vuli* rains. They are especially favored in Kwamsisi, where nearly everyone grows cowpeas. Cowpeas are also frequently grown during the *masika* rains in all but three highland locations (Vugiri, Kisiwani, and Kwadundwa). Green grams are grown in both *masika* and *vuli* rains, though a little less frequently than cowpeas. In the *vuli* rains they are grown everywhere except a few highland locations. Green gram planting is widespread in the *masika* rains as well in the eastern half of Tanga Region, excluding highland locations. Kwamgwe is an anomaly in all this; green grams were not mentioned by any farmer.

The reason cowpeas and green grams are found and beans are not found in certain villages has to do with their superior performance in drought-prone situations. As I have shown, the *vuli* rains are generally inferior to the *masika* rains, and although cowpeas and green grams may be in the field a few days longer than beans during the *vuli* rains, they have lower water requirements than beans. It appears that green grams and cowpeas mature a week to 10 days before beans during the *masika* rains.

TABLE 6.12. Planting Densities for Maize, Tanga Region

Village	Plants/ha (to nearest 100)	No. of triangles counted
Kwadundwa	31,300	8
Mgera	32,300	16
Kwediamba	39,400	18
Minazini	—	—
Kwamsisi	13,900	22
Mzundu	34,600	14
Kwamgwe	30,400	18
Mandera	12,200	19
Mlembule	14,200	10
Mkomazi	—	—
Vugiri	24,400[a]	2
Magoma	30,200	20
Kisiwani	27,400	12
Kiwanda	26,400	22
Daluni	19,900	14
Mwakijembe	11,000	23
Maranzara	23,600	23
Moa	9,200[a]	2
Average and total	24,000	243

[a]Only 2 triangles taken.

Spacing of Crop

One way to make a crop obtain a more reliable yield is to increase the amount of moisture available to each plant. If this cannot be done by irrigation, an alternative is to increase the spacing between plants. This gives roots a greater volume of soil from which to extract moisture. It appears that farmers in Tanga use a wider spacing than is recommended by Western agronomy. The standard spacing for a maize crop is 75 by 30 cm, which results in a plant density of 44,444/ha. The planting populations we found by counting crops in crop triangles showed a lower density. The spacing varied greatly from place to place, the lowest occurring in Mwakijembe, a dry area (11,000/ha). Table 6.12 gives the characteristic planting densities for maize in the transect villages.

The results of a simulation for 58 years at Mlingano of growing 150-day maize at two planting populations show that although one decreases yield with wider spacing, the reliability of getting a crop increases (fig. 6.10). In the example shown, the average yield was reduced from 16 quintals/hectare to 9 q/ha (a 44 percent reduction), but reliability was increased from

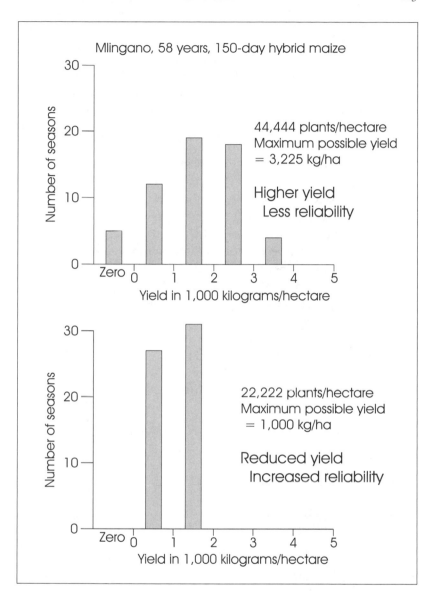

FIG. 6.10. Effect of using different planting populations.

49 percent to 93 percent. No seasons were total failures, compared with five total failures in 58 years with the higher plant population.

A simulation comparing the local practice in plant spacing with the standard 44,444/ha was done. Results are given in table 6.13 and figure 6.11. It must be stressed that the average planting populations (shown in table 6.12) mask the variability we found among the triangles. Farmers may use different spacings, depending on the soil and season and even on how well the rains have developed in the season prior to planting.

Figure 6.11, though difficult to read, summarizes the costs and benefits for farmers in using local average planting populations for maize. There is a "cloud" of village symbols (circles) in the lower right-hand corner, indicating that crops grown with the "local" plant population are more reliable (90 percent or more) than the standard 44,444 plants/ha. The exception is Vugiri, where the cost in yield is not worth the gain in reliability. In most of these villages, there is a loss in average yield, the price paid for greater reliability. The exceptions are Mwakijembe and Maranzara (Pongwe), where the lower plant population actually gives both much greater reliability and greater yield. In three other places the results are different and interesting. In Mgera, use of the local planting population increases yields substantially but improves reliability only slightly. In Mzundu, use of the local planting

TABLE 6.13. Comparison of Two Planting Density Strategies for Maize, Tanga Region

Village	Local spacing av. yield (kg/ha)	Av. percent of potential yield	44,444/ha spacing av. yield (kg/ha)	Av. percent of potential yield
Kwadundwa	—	—	—	—
Mgera	2,425	39	1,907	35
Kwediamba	—	—	—	—
Minazini	—	—	—	—
Kwamsisi	912	94	1,242	40
Mzundu	1,283	40	1,066	27
Kwamgwe	2,264	93	2,567	72
Mandera	1,056	100	2,143	55
Mlembule	1,204	98	1,582	41
Mkomazi	—	—	—	—
Vugiri[a]	1,322	100	5,863	97
Magoma	1,538	60	981	26
Kisiwani	1,862	97	2,078	66
Kiwanda	1,809	98	2,079	67
Daluni	1,569	94	1,512	41
Mwakijembe	703	92	5	1
Maranzara	1,414	89	897	30
Moa[a]	585	100	2,054	73

[a] Only 2 triangles taken.

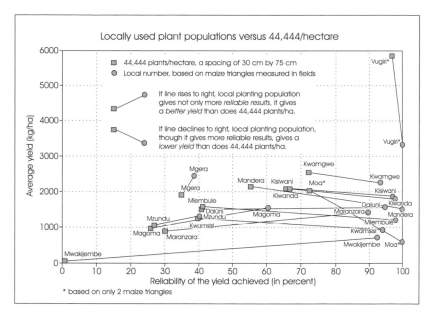

FIG. 6.11. Yield and reliability, maize, Tanga Region.

population improves both yield and reliability, but only marginally. In Magoma, use of local planting populations increases the average yield more than 50 percent, and reliability goes from 26 to 60 percent.

Farmers are frequently criticized by agricultural extension agents for using low planting populations in their fields. There is the argument that by spacing plants widely, one increases the evaporative/transpirative burden on the crop, since the air can circulate around the leaves of the plant with less hindrance. Many observers note that plants on the margins of fields, which are subjected to the fetch of dry advected air, are frequently more stressed and dried out than those in the interior of fields. Nonetheless, local farmers appear to have found that lower planting populations improve their chances of getting a yield, even if the amount may be diminished. As one farmer put it, "I like to space the plants out so that they have plenty of room to breathe."

In most villages, thinning the crop is one strategy nearly everyone uses when rains have failed; exceptions are Moa and Maranzara. Weeding more carefully is another common practice, although it is not much used in Moa, Maranzara, and Kiwanda.

Intercropping

Earlier I presented information on intercropping, or interplanting, in Mgera. Across Tanga Region, intercropping is common. About 80 percent of farmers interplant some crops. In most villages it is universal. It is less important in Moa, Mwakijembe, Magoma, and Mandera. The most common mixture is maize and cassava (fig. 5.5). In a compilation of 229 instances of intercropping, 150 of them involved maize and another crop, and there were 62 instances of maize intercropped with cassava. The cassava plant is small when the maize crop is growing, and when the maize has been harvested, the cassava, which takes longer to mature, is free to be in light and to use the moisture in the soil. The next most common combination is maize and cowpeas (28 instances). Some intercropping reflects the very different nature and dimensions of some crops. For example, perennials, such as bananas, cashews, coconuts, and fruit trees, are found in fields used for annual crops. Table 6.14 summarizes the main sets of combinations. Other crops mentioned in combination, in order of frequency, are pineapples, cucumbers, okra, tomatoes, black pepper, sugarcane, lemons, peanuts, cashews, jackfruit, cardamom, castor, finger millet, sorghum, spinach, and eggplant.

Respondents cited a wide variety of reasons for intercropping. Many of them were contextual, as for example the need to provide shade for cardamom and thus interplanting it with bananas or maize. Three reasons (summing to 79 of 136 explanations) stand out as most important: (1) it simplifies the work (*kurahisisha kazi*) and saves labor, especially the work of weeding; (2) many farmers have only one field at a time and live in areas with land pressure, and so intercropping maximizes a field's productivity; and (3) many farmers cite poor health or lack of energy as a reason for

TABLE 6.14. Crop Interplanting Combinations

Maize and cassava	62
Maize and cowpeas	28
Maize and beans	18
Maize and bananas	18
Maize and pumpkins/calabash	7
Cassava and cowpeas	13
Cassava and bananas	8
Cowpeas and green grams	5
Coconuts and bananas	10
Coconuts and oranges	9

intercropping. Two other explanations are that it helps in the timing of food availability (and the timing of labor) and that interplanted crops "do well together." Some farmers stated explicitly that intercropping enhances soil fertility. It was also mentioned as a way to control insect pests, reduce the labor of protecting against agricultural vermin, and reduce soil erosion. Finally, a number of people see intercropping as a food security strategy: "If one crop fails, another may not."

Provision of Supplemental Water

The crop does not care where its moisture comes from. Farmers can ensure adequate water for plants by growing them in an area with reliable, adequate rains; by preparing the seedbed so that water is impounded, filters into the soil, and does not run off; by weeding carefully; by mulching; and by using farmyard manure, which increases soil carbon and soil moisture-holding capacity. Still another option is to irrigate or provide water in other ways. When seedlings are small, perhaps even in a nursery before being set out, farmers may hand sprinkle the crop. Supplemental irrigation is used in higher areas, especially for cultivation of vegetables (Vugiri, Kiwanda, Mgera, and Kwamgwe), but is little used in lowland Tanga.[10] Irrigation is the basis for agriculture in Mkomazi, where rain-fed agriculture is all but impossible. The use of *mabonde* (low-lying areas) is a more common way to get more moisture to the crop. A majority of respondents in 12 of the villages plant in *mabonde* as a means of coping with a season of poor rains. It is simpler to list places where *mabonde* is not greatly used: Maranzara, which has poor sandy soils, Kisiwani, Moa, Mwakijembe, and Mlembule. In Mwakijembe, a few farmers plant in low areas along the river that are flooded at times, but no true irrigation is practiced. There are possibilities for increased irrigation in Tanga Region. One study stated that there is scope for increasing irrigation in several lowland zones (United Republic of Tanzania 1976: 59–67). In the Mkomazi Valley, crop productivity could be increased by growing maize or beans and pulses (green grams and black grams) under irrigation, rather than rice, which gives inferior yields. The Lwengera Valley, downstream from our study village of Magoma, has potentially irrigable land, although problems of soil salinity and alluvial soils that are heavy and hard to work constitute problems. Other potentially irrigable areas (lower Msangasi River and upper Umba River floodplain) lie outside the zone of this study.

Summary and Analysis

The simulation model has enabled us to explore in detail some of the environmental differences across Tanga Region and how people in different villages devise agricultural practices to use their local climate. We explored the timing of agricultural activities, which must take into consideration both weather and labor demands. We showed the results when one plants late and when one uses a different planting population. Across Tanga Region, farmers choose different crops, intermix them in fields in different ways, follow different schedules for planting and weeding, and space their crops differently. This variety in management reflects the extensive local knowledge farmers have of the area where they live—their sense of local conditions and possibilities. Although some farmers could do better in timing their planting and in their choice of crop or crop variety to grow, overall Tanga's farmers show a broad, informed, practical knowledge of crop requirements—which crops to grow and where, how, and when to plant them. They are alive to the issues considered in this chapter. To emphasize my conclusion once again: In the analyses presented in this chapter, in most cases local practice either *matched* the recommendations of Western agronomists, *beat* the recommendations of Western agronomists, or *maximized* some goal of local farmers, such as yield reliability or coping with labor bottlenecks, that may have been overlooked by Western agronomists.

CHAPTER SEVEN

Drought, Food Shortages, and Ways of Coping

In this chapter I explore ways in which people and institutions in Tanga Region have dealt with droughts and other environmental calamities that have affected the food supply. I begin with a sampling of the sorts of events, both natural and human created, that the people of Tanga Region have had to cope with. There follows a vignette of one famous drought, and I then explore various dimensions of food shortages and agricultural and attitudinal change from the time of Tanzania's independence (1961) to 1993. I give special attention to what people do when food shortages threaten the community.

When Things Go Wrong

As has already been noted, Handeni District, and Tanga Region more generally, have a long history of food shortages and associated disruption of livelihood. Giblin cites famine conditions in 1894–1896, 1898–1900, 1907–1908, 1910, 1916–1918, 1925, and 1932–1935, with severe local food shortages in years in between (Giblin 1990: 74). Underlying the story of historical change in Tanga Region, like an ominous pedal point, is an unending string of calamities that contribute to or reflect food shortages, brought on either by nature or by people themselves. I give some sense of the range of difficulties in the two brief litanies that follow. These litanies are quotations from archival sources that reflect the sorts of trouble that occur, with **bold** highlighting the nature or source of the calamity.[1]

First Litany

A small outbreak of **anthrax** occurred in the Digo North area during the year, and another of **heartwater** in the Bondei West area.... Early in the year it was discovered that **rinderpest** had caused a considerable mortality in the Digo cattle.... **Locusts** with their concomitant eggs and hoppers kept the Bondei and, to a lesser extent the Digo, as well as the Administrative and Agricultural staff, working at high pressure for the first six months.... Plagues of **army-worm** over some parts of the district have aggravated the position, as whole shambas [fields] of sorghum and maize have been wiped out.... [T]he Wazigua have a very genuine difficulty. If they plant their cassava shambas near to their houses they get **destroyed by cattle, sheep and goats**. If they plant them well away they fight a continual losing battle against the **vermin**.... **Baboons and monkeys** are very numerous and are incredibly bold in their raids on native gardens.... Much of the maize in the Nguu shambas has been destroyed by the **persistent rains** of July and August, the Kimba shamba was again **flooded** by exceptional rain.... [V]arious **pests** have done much disappointing damage to the cotton crops.... Owing to drought conditions in the early part of the year a number of sisal estates had **difficulty in maintaining an adequate supply of water** for decorticating their leaves which were themselves affected by the drought.... [I]f the moisture content is over 14%, ... maize is liable to heat and turn **mouldy** and make an ideal breeding ground for **weevils**.

Second Litany

Food security is not simply a matter of the adequacy of the rains or nature's intrusions. It is a deeply complex matter involving economic and social conditions. My second litany highlights aspects that are economic and social.

Transport and isolation: Kwamsisi Chiefdom ... and Kwekivu Chiefdom ... have been **cut off by rain** and it was only on the third trip that a lorry was able to get through to Kwamsisi the other day, and 2 or 3 attempts to reach Kwekivu in the last week have failed, as **bridges have been washed away**.... February's **rains** there **were torrential** (the heaviest in living memory). The result was that **miles of road and some bridges were washed away**....

Poverty: At present it is in Kwamsisi Chiefdom that there is the greatest **physical distress** as a result of the famine, partly owing to the difficulty of getting sufficient food for sale to them and partly because the population are **very poor** and **cannot buy** sufficient to keep going. . . .

Labor constraints and diet: In many areas . . . people are still hungry as this "Weeding Season" is always a difficult time, even in normal years. The **eating of the roots of a wild shrub** called "mdudu" is at present universal. These quickly fill the eaters' bellies, without providing sustenance; and a lot of present **tummy-trouble** is, I believe, due to this [the writer is referring to *Thylachium africanum,* a shrub with tuberous roots]. As regards co-coanuts, here or elsewhere the principal pest is the **thief**. [I h]ave suggested **famine relief work be closed to all women**, except for five vacancies on each work for cases of real need. The **fall in the price of sisal** [led] all estates to increase production as much as possible. This, together with the **failure of the harvests**, resulted in far more labour than usual seeking estate employment. . . .

Waste and poor planning: In 1925 [t]he District Officer, Pangani, **threw 100 bags of purchased maize**, bought in anticipation of food scarcity, **into the river** because there were no purchasers. There was a **serious famine** throughout Handeni district.

Analysis of an Environmental Chronology for Tanga Region

Using mostly archival sources, we prepared a chronology of environmental difficulties and disasters in Tanga Region covering a span of 54 years, from 1922 to 1975 (Appendix D). The chronology summarizes for each year the state of affairs (food, weather, and livelihood). The summation was created from a close reading of a large number of files from district and provincial offices in Tanga Region or in the Tanzanian National Archives. It was supplemented by data in Sumra 1975a: 26–28. We found information on 45 of the 54 years. In most instances the locale or district is mentioned. The point is readily evident that although drought is a common event and causes food shortages, there are numerous other things that affect food supply—in particular, too much rain, badly timed rain, agricultural vermin (bushpigs, baboons, rats, elephants, etc.), locusts and other insects, as well as crop pathogens.

A content analysis of the chronology permits us to know the number of times an item was mentioned. Information for some years is available for several districts. Since conditions may have been good in one part of

Tanga Region while poor in another, good yields and food scarcity may be mentioned in the same year. Let us hear the good news first, remembering that we have a base of 45 years: good rains, 9 times; good or excellent crops, 10 times. Now for the bad news: food shortage or food scarcity, 34 times; famine, 9 times; poor or failed rains, 22 times; excessive rains, 9 times; drought, 8 times; rain badly distributed, 3 times; rains late, 3 times; poor yields or crop failure, 13 times; need to replant crops, 2 times; locusts or grasshoppers, 10 times; agricultural vermin, 4 times; insects (weevils, army worm), 3 times; animal disease, 2 times; animal death from drought, 1 time; other (fire, unpredictable weather conditions, great physical distress), 3 times.

In summary, Tanga Region is an area of great climatic and environmental uncertainty. Food shortages occurred somewhere in the region at least 34 times (and can be associated with 31 of the 45 years). Failure of the rains is only one of many things that can go wrong and reduce food supplies. Indeed, if we add up all the other environmental items (excluding failed rains and drought) we get 39 items. Thus, the unpredictability of livelihood has as much or more to do with other aspects of weather and environment as it does with low rainfall.

There is some tendency for people to remember recent droughts as being worse than those in the more distant past. Those interviewed in 1972 said that the droughts of 1970 and 1971 were the most serious, hardly mentioning the serious drought and famine of 1953. In 1993, people cited the droughts of 1990 and 1991, although they did remember the drought of 1974. Some 60 respondents recalled the 1974 drought as very serious. It affected all villages east of Mlembule. It was particularly bad in Korogwe District (Mlembule, Mandera, Vugiri, and Magoma), as well as far to the southeast in Kwamsisi. What happens when there is a crop failure?

The 1953 Handeni Famine

Strictly speaking, Handeni District experienced a severe food shortage in 1953, not a famine. Few, if any, people died of starvation. Randal Sadleir, the district commissioner, declared food supplies "dangerously scarce," thereby making the district eligible for interest-free loans from the provincial commissioner for food purchases and associated costs—storage, transport, etc. (Sadleir 1999: 124). His act classified the situation as a famine. In subsequent discussion, the event came to be known as the 1953 Handeni famine.

TABLE 7.1. Maize Exports from Handeni District

Year	Exports in metric tons
1947	5,014
1948	7,216
1949	Nil
1950	5,412
1951	3,180
1952	566

The three years leading up to 1953 had not been good ones. The food situation in 1950 was precarious. Although rains were sufficient, there had been serious damage to field crops by army worms (*Spodoptera exempta*), particularly in the Kimbe, Mswaki, and Chanika areas of Handeni District. District Commissioner R. Thorne reported great physical distress as a result of famine in Kwamsisi Chiefdom.[2]

In 1951, half of Handeni District was facing a shortage at the close of the year and was dependent on maize imports. Handeni District had been the heaviest producer of maize and was called the "granary of the province." Table 7.1 shows the district's record of export sales. By 1952, surplus food stocks from good areas were seen as being barely sufficient to feed the people in the bad areas. There was no surplus to put in grain stores. As of October, food shortages had begun in Kwamsisi and Mazingara Chiefdoms. The stage was set for a food shortage to develop.

The *vuli* rains of 1952 generally failed (table 7.2). There was no surplus to put in grain stores. Food shortages had begun in southern Handeni and by October had spread to Chanika, Mswaki, and Magamba. Some 40,000 people were affected, and it would be a number of months before a harvest was available in 1953. The government had to act to ensure that food would be available.

The district commissioner, district officer, and others in the administration were aware of the potential for the development of food shortages. David Brokensha, district officer in Handeni District, 1952–1954, provided me with background information on the events in Handeni:

Sadleir was DC Handeni from May, 1951, and in early 1953 we had a succession of disasters. First, and most serious, was a food shortage, which Sadleir persuaded government to declare a famine—this was an important distinction, as once a famine had been officially declared, the affected district was eligible for famine relief, and maize could be distributed. We had good maize stores, and a generally adequate system of distribution, but there were two main problems. One, which

TABLE 7.2. Monthly Rainfall, Handeni Town (data in mm)

| Year | Masika | | | Apr. | May | June | July | Aug. | Vuli | Oct. | Nov. | Dec. |
	Jan.	Feb.	Mar.						Sept.			
1951	85	116	46	236	162	21	15	6	0	147	169	145
1952	144	111	96	44	73	12	1	8	16	7	50	26
1953	34	8	90	114	154	2	39	62	50	109	79	87

from your letter still exists today, concerned roads and transport. Lorries were not always available—the boma [district administrative offices] had one old 3 ton lorry which was very busy, and we hired lorries from Indian traders when we could. Even if we had the transport, the roads were not always passable.

The second problem was social rather than physical. We relied on the "Native Administration" to do the distributing for us, part of Indirect Rule. This meant that we used the chiefs, plus their usually rather rudimentary councils. One chief, Zumbe Salehe Saidi, of Mgambo Location, whose village, Kwankonje, was about 20 miles [southeast] of the boma, did not co-operate, but sold the famine relief supplies, or some of them, for his own gain. When we learnt of this, we investigated and were told by our network that the chief had said that if any-one tried to interfere with him, he would use his magical powers to make them mad, or very sick. Sadleir, undeterred, made up a case against the chief and persuaded the Governor to suspend the chief, pending a formal enquiry. This, by the way, was no easy matter, DC's alone had no power to suspend or remove a chief. (David Brokensha, pers. comm., 12 September 1993; see also Brokensha 1971)

On 31 January, Brokensha wrote to the provincial commissioner in Tanga as follows:

As expected the food shortage got worse during the hot dry weather in January and by the end of the month it had spread to a few areas in the Mswaki and Mgera Chiefdoms: only $1/4''$ of rain fell during the month, and most of the sorghum then planted has since dried up. An emergency meeting of the Ufungilo [council] was held from 26th to 28th January, to consider the situation.[3]

The railways were used to move grain in this period: "In January Railways moved 1,327 tons maize into Tanga from Northern Province. In January Railways moved 272 tons into Korogwe from Northern Province."[4]

By February 1953, Harry Collings, director of the Grain Storage Department, was sending letters and telegrams to the district superintendent of East African Railways and Harbours (EAR&H), Tanga, estimating how much grain had to be transported by rail and how serious the situation was:

1. I would confirm telephone conversation of the 16th instant, Harry Collings, when I drew your attention to the deterioration in the stock position at both Korogwe and Tanga. I told you that the stocks at both these stations were considered to be, if anything, too low, and that I could not consider them going any lower, the situation is that at both stations there is a little over a fortnight's requirements at the present rate of consumption only and that the present rate of consumption at Korogwe was very likely to increase if a severe food shortage, [that] threatened Handeni, did in fact take place.

2. The position, therefore, as I saw it, was that if we are to maintain the stocks at both these stations at their present level and we should endeavour to increase them, it meant that you must rail every month without fail a minimum of 1,800 tons, divided as to 1,000 tons to Tanga and 800 tons to Korogwe, and that this rate of railing would have to continue for at least the next 6 months. You informed me that you appreciated the position and that, if necessary, traffic would have to be diverted from other users, and I trust, therefore, we can look forward to an improvement in the railing for both these stations almost at once.[5]

As of early February, maize was being distributed in eight out of nine chiefdoms, through 62 centers, managed by African traders. This was true everywhere except Kwekivu—in the Nguu Mountains, west of Kwadundwa. Grain was also stockpiled in six areas that would later be inaccessible during the rains.[6]

At the end of February, a tone of desperation enters the correspondence, as in this telegram:

Subject: Tanga Province Maize Stocks

Tanga Township stock has been down to eighty tons this week which is less than three days requirements and present trickle of railments being eaten as fast as received. This critical position entirely due failure railways move supplies from Arusha and Moshi despite urgent daily requests. We have repeatedly advised you Tanga Township requirement is one thousand tons monthly and Korogwe 8 hundred tons monthly but February railments both stations has totalled less than one thousand tons being only half minimum requirement. I now confirm conversation yesterdate when you promised provide immediately emergency lift from Northern Province of at least 1000 tons. In view of very difficult food situation Tanga Province it is vital repeat vital you maintain railments at least 1000 tons Tanga and 1000 tons Korogwe monthly. Total requirement may increase. Essential you maintain steady flow railments immediately following present emergency lift.[7]

Brokensha reported on 4 March 1953:

The February rains failed (.32″) and any surviving early maize and sorghum dried up. Consumption of flour increased steadily, and there was a big demand for Relief Work. By the end of the month the position was serious, and if the present dry spell continues beyond 20th March, further assistance from Government will almost certainly be required. . . .

The remaining 85 tons in the Chanika Grain Store was exhausted by the 9th February, and a further 560 tons imported from Korogwe, making a total of 645 tons sold during the month as compared to 256 tons in January. At least 150 tons however has been stored in remote areas likely to be cut off in the rains. Distribution continued smoothly to 50 different centres and nowhere have stocks of food run out for more than a day or two.[8]

The administration was responding and appeared to have things under control.

Early in March, a new director of grain storage, G. H. Rulf, wrote to the traffic superintendent of EAR&H as follows:

This is to confirm your firm promise of yesterdate to rail 200 tons maize daily to Tanga and Korogwe from Moshi and Arusha divided, commencing immediately and continuing till further notice.

To prevent breakdown of supplies at Tanga and Korogwe it is imperative we build up reserve stocks repeat reserve stocks to at least 300 tons at both stations, this being one week's requirements.

Adequate supplies are available Moshi and Arusha and present critical position Tanga Province stocks is entirely due failure of railways move required tonnages particularly during February when Tanga railments from Moshi and Arusha were only approx. 1200 tons against requirement of 2000 tons.

Stocks at our Tanga and Korogwe stores were down to 80 and 90 tons respectively on 28 February and we have prevented breakdown of supply only by rushing supplies by sea from Mombasa. No further supplies remain at Mombasa and Tanga Province must now be fed entirely from Northern Province down Tanga line.

Loadings/railments for seven days 26 February to 4 March inclusive have been 626 tons from Moshi and 106 tons from Arusha—barely 100 tons a day. No wagons were available yesterday, Wednesday 4 March.

It is imperative that there should be no breakdown or holdup in railings at rate of 200 tons *daily* and tonnage must be divided as evenly as possible between Moshi and Arusha.[9]

In March, the provincial commissioner requested that District Assistant Reid and Crop Supervisor Scott be posted to Handeni for three weeks to aid in relief work, and later on they were so posted. A week later, a Mr. Kenyon was seconded from Lushoto to Handeni to help out. By mid-March, over 2,000 people were employed on "famine" relief works. Between 1 January and 31 March, 1,340 tons of maize were distributed in Handeni District.[10]

The corner in the crisis was turned in April:

In the first days of April heavy rain has fallen throughout the District, and the maize plantings are now well established. Some areas had rain at the beginning of March and this has now provided pigeon-pea, beans, cassava leaves and pumpkins to supplement the food supplies. Planting of maize has been on a big scale. Orders have been given for every house-holder to plant one acre of cassava before the end of April.[11]

Great advances have been made in establishing famine reserve foods in the planting of some 8,000 acres of cassava during the long rains. Another 10,000 acres under the village communal plantation scheme are under cultivation awaiting the forthcoming rains. Prospects for future periods of drought are therefore brighter.[12]

The effects of the famine were still occasion for comment in 1954: "The legacy from the 1953 famine continued to make itself felt throughout much of the year in almost every phase of life." "The year commenced with a continuation of the 1953 food shortage in the Nguu Zumbeates of Mgera and Kwekivu."[13]

How to prepare for food shortages has been a long and persistent problem in Handeni District. In the 1940s there was extended discussion among colonial officers about how to store grain: "Dryness of the grain is of prime importance, particularly with maize, and the latter should not be put in store if the moisture content is over 14%.... In Kenya the export of maize with a moisture content of over 12.5% is prohibited."[14]

Again, in 1950, before the food shortage discussed above, the district commissioner, R. Thorne, devoted much time to promoting the construction of strategically located grain reserves across Handeni District. His rationale follows: "In 1949 the crops were an almost total failure, and during the last few months the wealth of the people and the full resources of the Native Treasury have been used up in meeting the famine and in paying for imported staple at a cost to the buyer of between 3 and 4 times the price received by the seller per kilo in 1948."[15] He proposed building 15 100-ton grain stores in 1951, using a government loan, rather than stretching the construction out over a couple years. Provincial Commissioner M. C. Watt approved the project in September 1950.[16] Thorne then checked out the design of the stores with the Grain Storage Department, Moshi, to ensure that it was good and learned that the design for the stores was "entirely suitable." He then approached the Lushoto Native Treasury for a loan so that the work could begin.[17] There was a side discussion of switching to underground storage for some of the later facilities to be constructed, but this idea was abandoned because of the problem of termites.

In the end, three grain stores were built and two of them were filled with grain, courtesy of the Grain Storage Department. The ongoing task of increasing storage capacity with new go-downs (warehouses) was taken up by District Commissioner T. R. Sadleir.[18]

Institutionalized Coping with Food Shortages

As we have seen, the files in the district office, Handeni, contain ample evidence of historic food shortages in Handeni District. Other districts maintained similar files. The district agricultural officer and the district officer for Handeni kept a constant vigil over how much it had rained, how the crops were doing, and the status of food supplies and food prices in local markets. All this constituted a quasi-formal qualitative early warning system as to the risk that food shortages might develop and possibly deepen into actual famine conditions. The agricultural officer in every one of the 55 districts of Tanganyika had a similar responsibility (Porter 1979: 77; Fuggles-Couchman 1964). Countermeasures could be taken; and indeed there were formal directions from the highest level of the colonial government spelling out what should be done if the threat of food shortages developed.

The government's position on famine and famine relief was set out in Government Circular no. 12 of 1949. The circular instructed provincial officers to

1. Require adequate cultivation of drought-resistant crops known as being locally suited to the area (frequently, cassava and sweet potatoes).

2. Establish communal food reserves, including government-built and -run grain storage facilities, but also grain storage facilities built and managed by native authorities.

3. Prohibit export of foodstuffs when there was reason to fear that a food shortage might develop.

4. Report annually on the status of harvests and food supplies. District commissioners were required in their annual report, sent to provincial commissioners, to classify food supplies as "(i) plentiful, (ii) adequate, (iii) scarce, and (iv) dangerously scarce." If supplies were scarce or dangerously scarce, the district officer, accompanied if possible by the agricultural officer, was to do a careful local survey in person.

5. Prepare an estimate of the quantity of foodstuffs required to be issued in relief. If food was scarce or dangerously scarce, the district officer was to prepare this estimate.[19]

This information was conveyed to the chief secretary, the director of agriculture, and the director of grain storage; but bulk purchases of foodstuffs could not be made without permission except in a case of sudden emergency. The report was supposed to include full details as to areas affected, numbers of people involved, and local availability and locations of relief supplies. Estimates of the quantities of foodstuffs required from outside the province (or district) were required, as well as information as to the state of local native treasuries.

The purpose of such relief was "to prevent avoidable suffering and loss of life," but the circular insisted that this did not justify "slip-shod accountancy." It stated that *free issue of relief must in all cases be strictly limited to those who have no means to pay and are prevented by age, sickness, or other incapacity from earning the means to pay.*[20] All others were required to pay in cash at the time of receipt of food issues. An alternative to cash payment was "labour on relief works."

The circular observes that relief works can seldom be improvised satis-factorily, so provincial commissioners "are assumed to maintain an annu-ally revised list of permanent productive works which can be undertaken in their Provinces." It noted that such projects should be located close to those asked to work on them and that they not prevent cultivators from "preparing for next season's planting at the appropriate time."[21]

The provincial commissioners were expected and authorized to em-ploy (and pay wages to) local personnel for auxiliary famine relief. It was expected that all costs (administrative, purchase of foodstuffs, storage, constructions of grain stores, transport, etc.) would be paid from native treasury funds. The government would supply interest-free loans as nec-essary upon application from provincial commissioners. Officers in charge of famine relief operations were required to submit monthly reports to the provincial commissioners listing all expenditures.

After each food shortage had concluded, the provincial commissioner had to make a full report on the extent of the famine, its causes, numbers affected, and costs incurred. If costs of the food relief efforts reduced the native treasury to a level less than 25 percent (working balance and reserve funds) of estimates of current revenues, the government would provide financial help.

The circular ends by emphasizing that dealing with famine is a local responsibility of native authorities, and failure to respond by "every means in their power . . . will be regarded as culpable negligence and will be dealt with as such."[22]

In current times a highly sophisticated crop-monitoring and early warn-ing system called FEWS NET (Famine Early Warning Systems Network) has been created to perform the same task for a number of vulnerable countries in Africa (Frere and Popov 1979; Gommes 1985; FEWS NET 2003). It has been developed for the Sahelian states of Africa, plus an-other set of countries extending from Eritrea and Ethiopia in the north to Zimbabwe and Mozambique in the south (FEWS NET 2003). FEWS NET is funded by the US Agency for International Development, with support from the National Aeronautics and Space Administration, the US Geolog-ical Survey, and the National Oceanic and Atmospheric Administration, as well as input from local governments and a network of observation and reporting stations. It uses satellite imagery to assess the state of crop water availability and issues monthly food security reports for each country. It classifies its reports as emergencies, warnings, and watches, depending on the seriousness of the situation. In addition, FEWS tracks such aspects as food aid, distribution of seeds and fertilizers, food crop marketing, and trends in food prices.

The Tanzania Food Security Report for 14 November 2003 noted that the most recent Rapid Vulnerability Analysis for Tanga Region recommended releases of maize from the Strategic Grain Reserve in the following amounts: Lushoto, 218 metric tons (t); Korogwe, 203 t; Pangani, 358 t; Handeni, 702 t; and Kilindi (the new district created out of the western half of the old Handeni District), 621 t (FEWS NET 2003: 5). In FEWS NET reports in February and March 2004, there was still persistent food insecurity despite recent rains, and 280,907 people in Tanga Region were described as "food insecure" and in need of food aid (FEWS NET 2004a, 2004b: 3).

Strategies Adopted by Farmers to Cope with Drought and Food Shortages

In 1972, we asked about 24 separate actions farmers and their families might take when there is a drought and when the rains fail (table 7.3). The most common responses of farmers were, on the one hand, practical (buy food, 84 percent) and, on the other, spiritual (pray for rain, 77 percent). Three things done by more than half the farmers were to plant drought-resistant crops (56 percent), cultivate *mabonde* (low, wet places; 54 percent), and thin the crop standing in the field (51 percent), leaving fewer plants to compete for the remaining soil moisture. This latter response is somewhat surprising, because farmers commonly dislike thinning a crop.

Four practices done by more than a third of all farmers were to seek help from the District Council (43 percent; in colonial times this was called the Native Authority), seek wage work (39 percent), plant more acreage (37 percent), and weed more (35 percent). Almost a third said that they tried to get help from kinfolk or sold some of their cattle (or small stock).

The least used practices were regarded either as largely ineffective or measures that are rarely taken. Examples of the former are moving cattle (13 percent) and cultivating ridges (*matuta*, 12 percent). Practices engaged in by about a fifth of the farmers were to plant more seeds per hole (there is no contradiction between planting more seeds per hole and thinning, since these tasks occur at different times), irrigate, hunt and fish, and collect bush foods.[23] About an equal number ceased cultivation, since it was likely to be work expended with no reward.

TABLE 7.3. Percentage of Farmers Mentioning a Specific Action

	1972	1993
1 Ask help of a kinsman	31.1	47.7
2 Go live with a kinsman	8.4	7.6
3 Send children to live with a kinsman	6.7	8.2
4 Move cattle far away	12.6	17.1
5 Sell some cattle	30.7	49.1
6 Cultivate *mabonde*	54.2	60.2
7 Plant a drought-resistant crop	55.5	87.9
8 Plant more seeds per hole	29.4	10.5
9 Plant more acreage	37.0	50.9
10 Thin out the crop standing in the fields	51.3	82.7
11 Weed more	34.5	77.6
12 Irrigate	19.3	29.1
13 Cultivate ridges	12.0	31.8
14 Stop tending fields	18.1	5.4
15 Pray	76.9	91.9
16 Pay a rainmaker	26.5	11.0
17 Buy food	84.0	96.5
18 Hunt or fish	23.2	46.5
19 Collect food in the bush	20.2	34.9
20 Try to get a loan	9.7	34.9
21 Ask for help from District Council	43.4	35.1
22 Ask for help from cooperative society	7.6	16.4
23 Seek wage work	39.1	34.7
24 Do nothing	3.4	6.0

Examples of extreme measures are moving to live with kin (8 percent), sending the children to live with kin (7 percent), and trying to get a loan (10 percent). Only 3 percent said that they did nothing when faced with drought.

We studied the responses in detail, village by village, to see if there are regional differences among the 18 villages in their use or nonuse of particular ways of coping with drought. There is no strong pattern except in a few cases. Livestock sales are obviously restricted to those who own livestock. Gathering of bush foods is done more in western Handeni; and the average number of practices adopted is higher in Handeni than in the area around Korogwe (fig. 7.1).

In 1993, farmers in most villages did more of the 23 things (I exclude the "Do nothing" response) we asked about than they did in 1972 (table 7.4). The overall average of practices adopted in 1972 was 7.3; in 1993, it was 9.8. As in the past, nearly everyone buys food in time of food shortage and prays for rain when there is a drought. There was also a significant increase in seeking a loan. The largest increases were in choosing drought-resistant

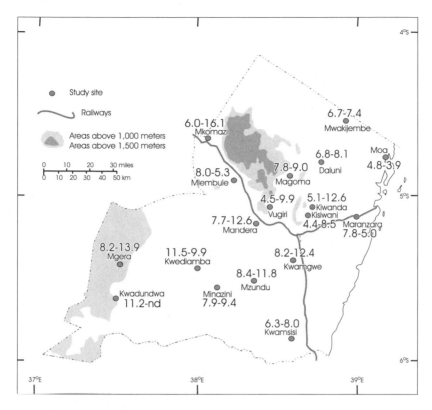

FIG. 7.1. Average number of actions taken by individual farmer in coping with drought. First number is for 1972; second number is for 1993.

crops to plant, planting a larger area, thinning, weeding, and cultivating ridges. There was also greater resort to hunting, fishing, and collecting bush foods.

There was a notable decrease in paying a rainmaker to perform rain-making rituals. There is a complex story here of cultural conflict. In Mlembule in 1972, one old farmer told us: "Since government has done away with the chiefs, the rainfall is all uncontrolled. The chiefs used to have power over the rain. I didn't experience famines in my youth. Now I do; there have been a lot more famines lately. Mr. Christopher [a European manager of a sisal estate near Mlembule] used to provide a white sheep for sacrifice." Feierman attributed this change to the effects of the national po-litical party (Chama cha Mapunduzi), which disapproved of using funds for

TABLE 7.4. Average Number of Actions Farmers Took

Village	1972	1993
Kwadundwa	11.2	—
Mgera	8.2	13.9
Kwediamba	11.5	9.9
Minazini	7.9	9.4
Kwamsisi	6.3	8.0
Mzundu	8.0	11.8
Kwamgwe	8.2	12.4
Mandera	7.7	12.6
Mlembule	8.0	5.3
Mkomazi	6.0	16.1
Vugiri	4.5	9.9
Magoma	7.8	9.0
Kisiwani	4.4	8.5
Kiwanda	5.1	12.6
Daluni	6.8	8.1
Mwakijembe	6.7	7.4
Maranzara	7.8	5.0
Moa	4.8	3.9

such purposes and saw traditional rainmaking ceremonies as a challenge to CCM authority (Feierman 1990: 245ff.).

The inevitable conclusion of our analysis of drought-coping practices is that people in Tanga Region continue to live in a chancy environment. Over the past two decades, however, they have adopted more practices, particularly with respect to selection and management of crops and fields, to reduce the risk in their livelihood.

Farmer Attitudes toward Nature

Both in 1972 and in 1993 we read each person interviewed a story whose purpose was to gauge the attitude the individual had toward nature (Douglas 1986).[24] The person about to read or tell the story was admonished as follows:

> This next question tries to find out farmers' attitudes toward nature. Be serious when you ask the farmer to reply to the question after you have read the story. Even though it is a story, it is not a foolish one, and the farmer's answer will be of great interest. Be sure that you read the story aloud without suggesting, by

the way you read it or by the tone of your voice, which man *you* would agree
with!

Here is the story:

> Once upon a time there was a great drought season and it went on for nearly
> a year. It got so bad that people had almost no food and they had to kill their
> livestock for food, and some older people even died because of the famine. The
> people gathered in a large *baraza* to discuss the famine and the rains. Many
> people spoke, but the people listened especially to three men.
>
> The first man said that there is little they could do about this drought for
> it was the will of Allah (God). He said the winds, the rain, the sun—all nature
> is powerful and man is very weak in comparison. They must suffer whatever
> happens and hope that God will be favorable next year.
>
> The second man said that what was wrong was that the people didn't have a
> good kind of agriculture. He said that men should try to understand the weather
> and crops; they should try to understand nature and live with nature as a partner.
>
> The third man said that people in other parts of the world didn't have the
> problems they were having here because they had solved them. These people in
> other parts of the world believed they were able to control the land and water
> and make it possible to grow crops even if the rains were poor. They felt they
> were superior to nature and could control nature.
>
> If the *baraza* had been held here during such a famine and you had been at
> the *baraza,* which of the three men would you have agreed with?
> First man___ Second man___Third man___

The positions presented in the story place humankind in three relationships with nature: subservient, cooperative, and dominant. The first has
an element of dependence and fatalism in the face of nature's power. The
second is the most ecologically and environmentally sanctioned, in that
people are viewed as being in a partnership with nature, a cooperation
based on an understanding of nature. The third has technical and scientific hubris: there is a technical solution to any environmental problem.
Table 7.5 shows the numbers of people who chose speakers 1, 2, and 3.
Overall, more people viewed nature as powerful and people relatively
helpless than either of the other two possibilities. In 1972, 48 percent
chose the first speaker; in 1993, 45 percent did so. The second speaker
(cooperation with nature) was chosen by more people in 1972 (33 percent) than in 1993 (25 percent). The third speaker was chosen by only

TABLE 7.5. Responses to the Story in the Eighteen Villages

	1972			1993		
Village	First man	Second man	Third man	First man	Second man	Third man
Kwadundwa	7	0	6	—	—	—
Mgera	5	6	1	5	0	4
Kwediamba	—	—	—	4	1	2
Minazini	—	—	—	6	6	0
Kwamsisi	13	6	1	9	8	3
Mzundu	10	4	0	7	1	2
Kwamgwe	4	6	2	2	3	3
Mandera	6	8	1	8	2	0
Mlembule	7	4	3	2	2	5
Mkomazi	2	5	5	3	0	4
Vugiri	2	10	0	0	0	9
Magoma	7	5	6	5	3	2
Kisiwani	4	7	3	6	4	0
Kiwanda	5	4	2	0	0	8
Daluni	8	2	6	5	6	3
Mwakijembe	9	1	2	3	4	3
Maranzara	10	1	1	8	2	1
Moa	8	3	3	6	1	4
Total	107	72	42	79	43	53
Percentage	48	33	19	45	25	30

19 percent in 1972 but by 30 percent in 1993; this is the most striking change over the two decades. The distribution of responses within villages is of interest, since some villages responded very differently. Because the numbers are relatively small, we are reluctant to read very much into particular sets of responses. In 1993, those interviewed in Vugiri and Kiwanda agreed overwhelmingly with the third speaker. These are among the richest and environmentally best endowed sites. There was relatively little change between 1972 and 1993 in Mzundu, Maranzara, and Moa; and respondents in these places tended to view nature as powerful and dominant.

We also examined the responses of 97 people interviewed both in 1972 and in 1993 to see if they had changed their views (realizing, of course, that they were all 21 years older). Thirty-one gave the same answer 21 years later, whereas 58 agreed with a different speaker. There was no consistent pattern—say, people moving from speaker 1 to speaker 2—although many now agreed with the third speaker.

We also inspected the response pattern (between both dates) among those who at the time of the 1972 research had been identified as one of the following: richest farmer, *balozi* (ten-cell leader), best farmer, poorest farmer, oldest farmer, or youngest farmer. No consistent pattern emerges among the 32 respondents so identified. Young farmers (in 1972, $n = 9$) almost universally chose speaker 2 in 1972, and either speaker 2 or 3 in 1993. In 1972, 9 of the *mabalozi*, or ten-cell leaders ($n = 12$), chose speaker 1. Two decades later, 4 of them had changed and now agreed with speaker 2.

In summary, one can say that there is a widespread view among farmers that they are relatively helpless to control weather and affect rainfall. This is most true of areas that are drought prone and that experience the greatest climatic variability. It is in the better-endowed, wetter, higher areas where one encounters the view that science and technology can solve most problems related to livelihood; and over two decades (1972–1993), the view that technology could solve problems gained many more adherents. As we saw in the earlier discussion on ways of coping, farmers in 1993 resorted to more ways of reducing livelihood risk, particularly in selection and management of crops and fields, than did the farmers interviewed in 1972. This is a hopeful sign for the region's agricultural development.

Changes in Crops Grown

Inconstant weather and problems with agricultural vermin are continuing problems in Tanga Region. They can be considered as givens in working out livelihood strategies. The changes over the years in cropping patterns represent local-farmer responses to economic opportunities emerging in the local and national economy and also responses to technical changes in agriculture, as better plant materials adapted to local conditions and better methods of husbandry emerge. We consider next the changes observed in crops grown in the period 1961 to 1972 and then in the period 1972 to 1993.

Agricultural Change between 1961 and 1972

In the decade between independence and 1972, certain crops were abandoned and others taken up. The greatest net loss was in cotton cultivation, with 24 fewer farmers growing it. This represents a great diminution in a cash crop of minor importance, because it was grown by only 43 farmers (18 percent) in 1972. It may reflect a reaction to the departure of

British colonial officers, because under the British administration a great deal of effort was devoted to getting people to grow cotton (John Ainley, pers. comm., 10 June 2002; Ainley 2001: 52). The other major losers were cassava, abandoned by 19 farmers, and sorghum, abandoned by 9 farmers. But while cassava was abandoned in some areas, its cultivation increased in others, and it was grown everywhere except Mkomazi and Mlembule, where it is really dry. Over half of the farmers, up to 100 percent in some villages, grew cassava east of Korogwe to the coast.

The regional pattern was as follows. Cotton cultivation declined in Minazini and Kwamsisi, in central and southern Handeni District. Sorghum declined in importance in central and eastern Handeni District (Kwediamba, Mzundu, Kwamgwe, and Kwamsisi) and in southern Tanga (Muheza) District (Maranzara and Kisiwani).[25] The decline in cassava cultivation was especially noticeable in Magoma and Mlembule, both of which are in Korogwe District and associated with sisal estates.

On the other hand, cassava cultivation increased in Tanga District itself. Seven farmers in the six villages in Tanga District began to grow cassava in the ten years preceding 1972, whereas before that they had not grown the crop. Seven may not sound like a large number but it constitutes 10 percent of those interviewed in Tanga District.

Certain crops made large gains during the decade and were grown by more farmers over wider areas. To some extent this represents the movement of a crop that was well established in one area into an area where it was less frequently grown. Among crops taken up as new crops by farmers, the most important were maize, coconuts, cashews, bananas, and rice. Maize experienced most of its expansion in eastern Handeni District and southern Tanga District, a northeastward movement. In central and western Handeni District more than 50 percent of farmers grew bananas. Banana cultivation was strongest in western Handeni, reaching 85 percent in Mgera, a very old, established center (its eastward diffusion has been recent). Banana cultivation declined to nearly zero at the coast. There was very little banana cultivation in Tanga District, except in Kiwanda and Kisiwani. In Handeni District in 1972 the banana plantations looked a bit forlorn—spindly, struggling, stressed, and made ragged by the wind—but the banana was nonetheless established widely as a food crop.

Coconuts increased in importance in central and northern Tanga Region, stretching from northeastern Handeni District and adjacent lowland Korogwe District to all of Tanga District, including Mwakijembe, far in the north.

Increased production of cashew nuts was a case of additional cultivation in situ. Cashews were added as a cash crop in the zone where they were already important, that is, within 40 km of the coast.

There was a slight increase in rice cultivation in its peer area (Mlembule and Mkomazi) and in southwestern Handeni District (Mgera). Decreases in rice cultivation in Tanga District were very slight.

In other cases, the agricultural change represented a new crop altogether. Cardamom (*iliki*) falls into this category, and while a few farmers may have been growing it before 1961, 24 farmers began to grow it in the decade 1961–1971. Most of the new cardamom cultivation occurred in the eastern Usambara Mountains (in Kisiwani, on the road up to Amani, and in Kiwanda, Daluni, Magoma, and Vugiri); but there was also some new cultivation of *iliki* in Kwadundwa in southwestern Handeni District.

Still other crops were taken up and set aside by farmers from time to time, but this was done by roughly equal numbers, so that the mix of crops overall and the appearance of the farming landscape did not change. For example, in the decade 1961–1971, cowpeas were abandoned by 11 farmers and taken up by 14; seed beans were taken up by 16 farmers, while another 14 stopped cultivating them.

We made an inventory or count of all crops mentioned by farmers in our sample. Some 50 crops were mentioned in all, some by most everyone (maize), a few only once (e.g., wheat, ginger). In all there were 1,900 citations of crops among 241 respondents, an average of about eight crops per farm family. (Even so, this probably undercounted the actual number of crops grown because chilies, calabashes, some garden vegetables, and various spices and herbs were probably overlooked when the farmers provided their answers.)

In 1972 we asked farmers to name their three most important cash crops and their three most important food crops. These were given a special scrutiny to see which emerged as lowland Tanga's key crops. On a second level we examined how frequently a crop was cited as being grown by a farmer (since we also tried to find out the names of all the crops the family grew) to determine whether it was among the top six or not. From this we identified a number of important secondary crops, grown by many but not necessarily counted among the main food or cash crops. The 2,083 citations consisted of 1,342 citations of crops that were among the top six crops, and 741 others mentioned as being grown.

The most important crops were, in order of precedence, maize, cassava, beans, cowpeas, bananas, and rice (table 7.6). These six crops account for

TABLE 7.6. Main Crops Grown in Lowland Tanga Region as Cited by Farmers Interviewed in 1972

A Crop	B Among top six (percent)	C Mentioned by farmer (percent)	D Difference C − B (percent)
Maize	94	97	3
Cassava	70	79	9
Beans	49	65	16
Cowpeas	49	66	17
Bananas	39	52	13
Rice	37	42	5
Coconuts	29	38	9
Cashew nuts	24	31	7
Green grams	22	41	19
Sorghum	17	29	12
Oranges	16	28	12
Cardamom	15	18	3
Tomatoes	13	28	15
Sweet potatoes	12	22	10
Mangoes	8	29	21
Spinach	1	18	19
Breadfruit	2	13	11
Okra	2	14	12
Cabbage	0	8	8
Cucumbers	0	6	6

61 percent of the citations of the most important crops. Maize was viewed by 94 percent of farmers as a key crop, followed by cassava at 70 percent, beans and cowpeas each at 49 percent, bananas at 39 percent, and rice at 37 percent. These crops were followed by coconuts, cashews, green grams, sorghum, oranges, and cardamom, accounting for another 22 percent of citations of important crops. Between 15 and 29 percent of farmers cited these crops as among their most important food and cash crops. Thus, a mere 12 crops, among 50, accounted for 83 percent of crops viewed by farmers as their most important food and cash crops.

When we examined the listing of crops by presence (rather than primacy), we found a general similarity, of course, with the list of key cash and staple food crops, but the list increased to 15. Table 7.6 shows that 97 percent of farmers in lowland Tanga grew maize (the exception was Moa, on the Indian Ocean coast, where few farmers grew maize). Almost 80 percent grew cassava, about two-thirds of farmers grew cowpeas and beans, and half grew bananas.

A measure of the "secondariness" of a crop is the difference between the crop viewed as a staple and the crop grown to some extent. Several crops were not regarded by farmers as key staples, but many farmers grew them in small amounts. The percentage shown in the column of difference (D) in table 7.6 shows this characteristic. Mangoes, cowpeas, green grams, bananas, beans (the preceding four are key staples in their own right for many of lowland Tanga's farmers), oranges, breadfruit, and sweet potatoes were regarded by many farmers as secondary crops. An assortment of vegetables rounds out the kitchen garden. Such crops as tomatoes, spinach, cabbage, okra, cucumbers, eggplant, pumpkins, green peppers, and chilies are grown by many farmers, but as supplements to the diet rather than as staples.

Agricultural Change between 1972 and 1993

In 1993, we sought to understand what changes had occurred over the two decades in crops grown and in crop yields. We asked: "What crops have you started growing and what crops have you stopped growing?" Ninety-four respondents said that they had given up at least one crop. The total number of crop citations numbered 168; 119 respondents started growing at least one new crop. The new crop starts numbered 273. Clearly, there was a lot of change and experimentation between 1972 and 1993. The main increases were in growing maize, cassava, rice, coconuts, and oranges and other tree crops.

We also asked: "Which crops do you grow more of or less of in 1993?" The number of crops being planted more was 210; and 78 percent of the increase was accounted for by only three crops: maize, cassava, and rice. Only eight people increased growing of pulses and oilseeds, or only 4 percent. As for decreases in planting, 145 people said that they planted less of a crop. The number of crops being planted less was 244, of which 58 were vegetables and 167 were pulses and oilseeds. This accounts for 92 percent of the decline. There has, in short, been a simplification of agriculture and perhaps an impoverishment of the diet.

To summarize, 119 people started growing maize, cassava, rice, and oranges, while 94 stopped growing cotton, cardamom, cashews, pulses, and oilseeds. Some 125 people grew more maize, cassava, and rice, while 145 people grew less cowpeas, beans, green grams, peanuts, and vegetables.

The reasons for these changes are complex. To some extent they are less a matter of trends than they are of the meteorological flux of seasons, some of which bring fortune while others bring misfortune. In explaining crop yield increases, a number of farmers pointed to better field management on their part (careful cultivation and weeding) and use of better plant material (hybrid and drought-tolerant species). By far the most cited items, however, were technological changes (42 instances) and timely and reliable rains (44 citations). The technological items included extension advice, access to a tractor, use of commercial fertilizer or farmyard manure, early or timely planting, and (in Vugiri) scientific agronomy. Other reasons cited for increased yields included luck, a larger family (with greater labor capacity), and "good crop prices."

On the other hand, reasons given for decreases in plantings featured adverse weather (60 citations); agricultural vermin, including insects and thieves (31 citations); tired or depleted soils (13 citations); poorer husbandry by farmers (13 citations); and a miscellany of crop diseases, "unreliable" crops, and low prices.

As noted in chapter 5, the most ecologically sound crops for Tanga Region are tree (or treelike) crops (coconut, citrus, mango, pawpaw, jackfruit, and cashew trees in the lowlands; bananas, coffee, tea, avocado, cardamom, and teak in the highlands; Hamilton 1989). The fact that more farmers are giving greater attention to tree crops that can be sold is an encouraging sign.

To return to the question of the possible impoverishment of the diet, is it possible that, even though farmers may have given up the cultivation of a number of pulses, oilseeds, and vegetables, they were purchasing them in the market? Such data as are available suggest that there has been little or no increase in the proportion of food production that is sold or purchased rather than produced and consumed within the household. Most indicators suggest decline—the loss of jobs in the sisal industry leading to lower sales of produce to workers, declines in the distribution of improved seeds, declines in the amounts sold to the Tanga Regional Cooperative Union (although this may reflect the low prices they pay), a 50 percent drop in the retail sale of sugar (table 7.7). The only indicator of agricultural improvement is a fourfold increase in the use of commercial fertilizers. A quotation can provide context: "The extent to which *food crops* are marketed is extremely difficult to estimate. This is so for two reasons. First, no data are available on the quantities which are exchanged on local markets. This fact is particularly unfortunate since there are a great number of wage-earning persons, working mainly on estates, who buy considerable amounts of staple foods, such as maize, cassava and beans on local markets. Second there

TABLE 7.7. Production, Sales, and Distributions in Tanga Region (all data are in metric tons)

Item	Early year[a]	Later year[b]
Sisal	90,235	28,076
Cotton	24	48
Tea	5,690	5,773
Arabica coffee	499	384
Cashew nut purchases	3,083	2,162
Distribution of improved seed		
Maize	386	29
Rice (paddy)	3	0
Sorghum and millet	12	0
Wheat	21	0
Beans	10	19
Distribution of fertilizer		
Sulfate of ammonia	99	400
Calcium ammonia nitrate	2	62
Triple super phosphate	203	52
Urea	0	749
Retail sale of sugar	7,491	3,873
Tanga Regional Cooperative Union purchases		
Maize	68,100	99
Rice (paddy)	6,400	46
Pulses (beans)	9,500	400

Sources: United Republic of Tanzania 1976; Bureau of Statistics 1992b.
[a] Early dates are 1974, 1977–1978, 1980–1981, 1982–1983.
[b] Later dates are 1989–1990, 1990, 1990–1991, 1991.

is, of course, no statistical information on the quantities passing through illegal market channels. For some crops it appears that sales through black markets are greater than those through official channels" (United Republic of Tanzania 1976: 105, italics in original).

Only a small portion of agricultural production is sold. "Less than 10% of the main staple foods, i.e., cassava, maize, bananas are marketed; the only exception to this rule appears to be beans" (United Republic of Tanzania 1976: 107). Most of the comparisons between years in table 7.7 suggest decline or stagnation in agricultural marketing.

Summary and Analysis

Food security has always been a constant concern of farmers and administrators in Tanga Region, and it continues to be today. So long as people farm in Tanga Region they will worry about the rains, the state of crops, and the adequacy of the food supply. When food shortages emerge (or threaten to do so), everyone is affected. Farmers try to save their crops, food prices rise as scarcity increases, and administrators at all levels of government try to avert the most serious outcomes—starvation and death—by ensuring that food relief supplies are available and distributed. We saw how this worked in 1953 in Handeni.

We found that farmers in 1993 were engaged in more practices designed to cope with drought and food shortages than they were in 1972. The largest increases in practices were agronomic: choosing drought-resistant or drought-escaping crops, planting larger areas, thinning, and weeding more carefully. In assessing farmer views about nature and their ability to control events affecting their livelihood, we found that there continued to be a high level of fatalism. One could not predict or control the rains or the outcomes in agriculture. For most, it was in the hands of Allah—Inshallah! (God willing). The exceptions appeared to be in the higher-potential villages (Vugiri and Kiwanda), where farmers were nearly unanimous in the view that with science and technology one can understand the environment correctly and make it reliably productive. The government itself seemed to have a modernist, technological view, if its opposition to farmers' paying rainmakers is any guide.

The changes noted in crops grown between 1961 and 1993 reflect the willingness of farmers to try different things—new crops and new ways. The major changes in the earlier period (1961–1972) were the abandonment of cotton, sorghum, and (in places) cassava and an increase in cultivation of maize, coconuts, cashews, bananas, rice, and cardamom. In the later period (1972–1993) maize, cassava, rice, and oranges experienced increases, whereas declines occurred in growing cotton, cardamom, cashews, pulses, and oilseeds. The encouraging feature of many of the increases was the emphasis on ecologically appropriate perennial tree crops in both lowlands and highlands. The cause for concern was the remarkable decline in the growing of nutritious pulses (beans, cowpeas, green grams, peanuts) and oilseeds (simsim, sunflower, and castor). All of these changes occurred in the presence of a growing population. The added people means that new markets have opened up for farmers to sell to. How the expansion of maize, cassava, and citrus and other tree crops since 1972 has affected (possibly impoverished) the diet of farmers and their families is impossible to say.

CHAPTER EIGHT

Local Knowledge, Sustainability Science

In this final chapter we turn to the question of the long-term prospects for Tanga Region and how sustainable livelihoods could be fashioned for its growing population. Could there be a transition to sustainable livelihoods? Are there necessary enabling preconditions? What potentially fruitful cooperation could there be between local farmers and scientific organizations that are concerned with crop production? Could the social capital embodied in the knowledge of the local residents and in their way of doing things be married to Western agronomic science (Mabogunje and Kates 2004)?

Each of us lives in a box; none of us is totally free and autonomous. We are all constrained by social, economic, and political laws and conventions. In thinking about agency and structure among the people of the 18 villages, I have noted that there were wealthy and powerful individuals and families, and there were poor and impoverished families. In any time period, all farm families in Tanga Region lived with constraint from some source, whether it was under the patronage of wealthier, more powerful individuals in the 1880s, the dictates of the *akidas* under German administration, the regulations set forth by the British administration, or the "top-down" *ujamaa* program of the 1970s.

Are we witnessing a transition to sustainable development in Tanga Region? Some comfort can be taken from the fact that secular trends in rates of population growth are downward. In the period 1957–1967, the annual rate of increase was 4.2, from 1967 to 1988 the rate was 2.5, from 1988 to 2002 it was 1.8. It may be that some of the decline between 1988 and 2002 was AIDS related. Whatever the case, we need to see how these people, in their greater numbers, have been doing.

We have seen in this study that the people of Tanga Region live in a difficult, chancy environment, particularly those lowland farmers dependent on rain-fed annual crops. The population of the region grew from 771,060 in 1967 to 1,642,015 in 2002, more than doubling in 35 years. In that time the region's major economic activity, sisal production, declined to almost nothing, taking with it employment opportunities and a market for agricultural produce. Tanga Town, the major urban place, experienced stagnation or decline in the use of its harbor and in commercial and industrial enterprises. (It should be remembered that, after 1924, Arusha and Moshi Districts, for which the Tanga railway was originally built, came to be served by a line connecting to Mombasa, Kenya [Hill 1962: 191]. Mombasa's port facilities at Kilindini are better than those of Tanga, and the trip to the coast is 86 km shorter. Thus, after 1925 much of the freight from the Kilimanjaro area was diverted away from Tanga Town.) The Tanzanian government's experiment with *ujamaa vijijini* in the 1970s occasioned major economic and social disruption and generated mistrust of the national government among local people.

We explored a number of measures of social and economic development and human well-being in Tanga Region as of 1972 and 1993. We found that although some villages had prospered, taken as a whole there had not been much improvement in material well-being; access to water was no easier in 1993 than it had been in 1972, and the health status of villagers did not appear to have improved. Indeed, the 2002 census results suggest that AIDS and HIV infections have made serious inroads. Further, in the face of major population increases, there has been a simplification in the diet, with greater emphasis on the cultivation of maize and cassava and a diminution in the importance of some really nutritious crops—pulses (beans, cowpeas, green grams, peanuts), oilseeds (simsim, sunflower, and castor), and a number of garden vegetables.

Probably the most important causal factor in Tanga's lack of economic progress has been the increased pressure of population on resources and the way the farmers have responded to this pressure. In 1965 Ester Boserup presented her influential, anti-Malthusian thesis on agricultural growth. She argued that agricultural innovation and increased productivity result from slow, sustained population pressure on agricultural systems, as they evolve from forest fallow, to bush fallow, to short fallow, to multicropping (Boserup 1965). Indeed, many instances have been documented of such intensification (Turner, Hyden, and Kates 1993; Maro 1974). A requisite for intensification is a resource base that will respond positively to inputs of labor and technology. There are high-potential areas in Tanga Region where Boserup's thesis could apply and indeed where intensification has been observed (Scheinman 1986; Heijnen and Kreysler 1971), but much of lowland Tanga is not blessed with a natural-resource base amenable to such intensification. Some improvement can probably be done in lowland Tanga by shifting emphasis from annual to perennial tree crops, by judicious use of drought-escaping and drought-tolerant crops, and by construction of bench terraces on hillsides as a means of increasing soil moisture.

Institutional Agricultural Research

Many economic and social investments must be made to bring about a sustainable transition. Huge investments in infrastructure (roads, clinics, schools, markets, agricultural machinery, cooperative societies, industrial employment opportunities, etc.) need to be made, yet agricultural research institutions also have an important role to play. For this reason I present here a discussion of institutional agricultural research as well as a brief account of personal experiences in understanding farmer knowledge and the workings of research institutes.

How can institutions play a useful role in improving the productivity of agriculture and the well-being of farmers? Paul Richards (1986: 156) cites the ironic appraisal of much Western policy research given by Robert Chambers: "Chambers (1983) argues that agricultural development policies are unduly affected by urban, roadside, dry-season, male-based perceptions of rural life and its problems." Richards further contrasts the mind-set of research institutions versus rural farmers. Institutional research is analytical; it breaks down the landscape into constituent parts (swamp, bottomland, slopes) and devises a "best use" for each terrain. "By contrast, farmers view the landscape in synthetic terms. The landscape is habitat. It is a living form" (Richards 1986: 156).

The major research institution in the midst of Tanga's farmers is the Mlingano Agricultural Research Institute (MARI). (*Mlingano* is a Kiswahili word meaning "likeness," that is, to make equal or to harmonize.)

It was started as a research station for the sisal industry but after independence became a more general research institute responsible to the entire nation for soil survey and soil nutrition. They issue annual reports on their research, as well as reports on particular research projects. Their reports are divided into six types: (1) soil and land resource inventories of regions and districts, (2) reconnaissance soil surveys, (3) detailed and semidetailed soil surveys, (4) site evaluations and soil appraisal studies, (5) soil fertility studies, and (6) miscellaneous publications (National Soil Service 1992a). MARI has much to contribute, through its soil fertility studies, to increasing crop yields among Tanga's farmers.

Next door to MARI is the Ministry of Agriculture Training Institute (MATI), which has a two-year diploma program as well as short courses for farmers, students, agricultural extension workers, and other professionals (Mshana and Mduma 2000: 186). This is clearly a resource that should be in the forefront of helping Tanga's farmers make their agriculture more productive, reliable, and ecologically sound.

MARI has historically been underfunded and has had to struggle to carry out its work. When we were there in March 1993, the government had fallen behind on paying the research station's electricity bill, and TANESCO, the electric company, had turned off power. As a stopgap measure, MARI had rented a generator. When the lease ended, the company that owned the generator came and took it away. As a consequence, MARI and MATI (which was also served by the generator) were both without electricity. No work could be done because computers could not be run, laboratory tests could not be performed, and cartography, printing, and photocopying could not be done. MATI could not offer courses. Furthermore, at MARI there were challenges in keeping and training a staff of high quality and getting the work done. At any time, some staff people were abroad training (which is good), but there was considerable staff turnover. In 1993, I examined the MARI research library. It seemed to me that most subscriptions to journals had stopped in 1973.

To ask that research institutions work closely with local people may be asking for something nearly impossible. The mandate for national institutions is to respond to requests for research and service nationwide. A study of the many soil surveys and site evaluations done by MARI shows that 93 percent were undertaken outside Tanga Region or, if done within Tanga Region, were performed on sisal or tea estates (National Soil Service 1992a). There were only four studies done in Tanga Region that concerned local villages.

Personal Experiences Concerning Local Knowledge and Research Institutions

I have had humbling experiences as to how much I miss when looking at a rural landscape. I know that my inquiries can never achieve complete knowledge of what Tanga's farmers know and why they do what they do. This was vividly brought home to me through an incident during fieldwork in Kimutwa (Machakos District, Kenya) in 1979. A farmer and I were talking, touring his fields. He paused, pointed at some beans, and said: "Look at that." I saw nothing. He then pointed out to me a narrow swath through the beans where the leaves had slightly less turgor and greenness. "Come look," he said, and showed me the undersides of the bean leaves. Rust-colored spores of a pathogen covered the undersides. Then he pointed to a nearby hillslope. The source of the spores, he said, was up there, and the beans had become infected at night when there was cold air drainage. This is one of the myriad things that farmers know and that visiting scholars are likely to know only if told.

We made a substantial effort, devoting days of research time, in an attempt to find records at MARI of the details of field trials for various crops.[1] The purpose was to develop a data set that included the yields the trials resulted in and various phenological details, including most importantly date of planting and date of harvest. Ideally, we wanted also to have date of physiological maturity and date of flowering. Further, we needed to know where the trial was conducted and if possible the soil and slope of the experiment site. Our purpose in collecting these data was to use the yields and phenological dates to calibrate the model that examines energy-water-crop relations (Appendix A).

Our effort was inconclusive, and after much effort a paltry set of yields and phenologies was gathered. It appears that agricultural research institutions resemble in many ways academic departments at research universities. That is, they reflect the freedom, anarchy, and individualism of faculty, and they do not have good institutional memories. The agronomists and other scientists designed their studies, saw that they were carried out, and then reported them in the institute's annual reports (National Soil Service 1988a, 1989a, 1990, 1992b). If the report included yields and planting dates but omitted how long the crop grew or the date it was harvested, one was out of luck. When the scientist's contract ended, he or she was transferred elsewhere, and the research notes got packed away or tossed out as the scientist departed.

Ostensibly, there was a book recording the results of trials, kept in the Soil Fertility and Management Section. Each trial had a research officer and a field officer assigned to it. The former was responsible for design, supervision, and analysis; the latter was responsible for the actual conduct of the experiment (overseeing weeding, fertilizer applications, harvest, etc.). The fact that two people were responsible no doubt allowed some data (of interest to me, at least) to fall through the cracks. Despite the best efforts of Ms. Karumuna, head of the section, we were never able to find this "holy grail." In a study of 178 individual trials, whose results are contained in the National Soil Service *Progress Reports* cited above, we found that something was missing in all but 16 trials—either yield, harvest date, or planting date. But even these 16 trials were divided among several crops (maize, cowpeas, and pigeon peas), six places (Mlingano, Handeni, Korogwe, Kwahulonge, Segera, and Suluti), and different rainy seasons (*vuli* and *masika*). The largest number of trials of one crop in one place and one season was five—maize grown at MARI during the *masika* rains (1988, 1989, 1990, 1991, and 1992). Even these five trials, part of a long-term experiment on the effect of nitrogen and phosphorus, had different fertilizer applications, and so any resultant yield variations caused by drought were indeterminate. In the end, I did not have sufficient data to calibrate the energy-water-crop model, and we have only the calibration I used in the original application to Utafiti, the Tanzanian Commission for Science and Technology (fig. 8.1).[2] This calibration was based on 33 trials conducted in Machakos and Kitui Districts, Kenya, that were part of John Corbett's dissertation research (1990).

When Institutions Try to Reach the Local People

Research institutions find it hard to incorporate local knowledge. At the International Institute of Tropical Agriculture (IITA), Ibadan, Nigeria (one of the Consultative Group on International Agricultural Research [CGIAR] institutions), they had, in the mid-1980s, a model village, with associated fields, the idea being that researchers could see how people farmed and learn about local agricultural practice. Yet at the end of each day the members of the village were transported back to their real villages outside the IITA compound (which was surrounded by a 7-foot-high Cyclone fence topped with barbed wire, with an adjacent roadway that was constantly patrolled).

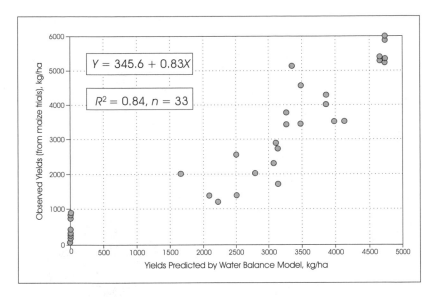

FIG. 8.1. Water-balance model verification for maize trials in Katumani, Kitui, and Itumba, Kenya. (Source: Corbett 1990: 60)

IITA had another experiment, on maize and cowpeas, to study the effect of different technical inputs on crop yields (fig. 8.2). These treatments were to be compared with the indigenous slash-and-burn method, for which they engaged a local farmer. The man engaged to clear and manage the plot was to use his own tools and was told to do things exactly as he would have back in his village. He must have thought to himself, "Oh boy, if I do a good job with this field, maybe I will get a permanent job at IITA," so he worked very hard on his plot. In any event, his plot had the highest yield of all of the experimental combinations.

Staff members at these organizations know when their efforts have not been successful. MATI in particular has an extension, or outreach, function and should be diffusing useful practices and technologies. For example, one of their programs deals with animal traction (donkeys and young, castrated oxen). One proposed use of traction animals is to pull cultivators between row crops, such as maize, to reduce the work and improve the quality of weeding and to overcome serious labor bottlenecks that occur at weeding time. Despite extension efforts, only five farmers within a radius of 20 km of MATI used draft animals for weeding in 2000 (Mshana and Mduma 2000: 186).

FIG. 8.2. Dunstan Spencer, an agricultural economist, discusses a sign showing the layout of a cropping-systems trial at the International Institute for Tropical Agriculture, Ibadan, Nigeria. In a field trial of maize and cowpeas, the traditional cultivation method was compared with a series of experiments that added successive elements: recommended spacing, improved variety, improved weed control, *Leucaena* alley (a way of fixing and adding nitrogen), and fertilizer. Photo taken in August 1986.

The adoption of animal traction is complicated by human and economic factors. In the first place, farmers and their animals have to be trained in the use of equipment. Second, the equipment itself is not cheap or even readily available. A disc plow cost T Sh 359,000/- in 1989 (about $2,300), a substantial sum for the typical farm family (Scheinman 1989: 3). The purchase and use of a tractor was even more problematical, since farms are small and fields are not generally laid out in ways amenable to tractor plowing. Tractors are such large capital investments that they are affordable only if shared among a large number of users. A Valmet 604 tractor cost T Sh 2,995,000/- in 1989 (about $19,300). There may be still deeper reasons why animal traction and mechanization have not caught on among farmers in Tanga Region. The hoe (*jembe*) is the standard, universal agricultural implement—inexpensive and readily available. Stuhlman, the German who established the Amani Agricultural Research Station in 1902, felt that hoe cultivation was not an inferior technology, but a different one (Koponen 1994, cited in Sosovele 1999: 253). It would be most useful for researchers and agricultural extension personnel to work closely with local farm women (who do most of the weeding) and men to find out what the constraints on mechanization are and what might be done to remove them.

Plant breeding done in the laboratory without consultation with local farmers and then transferred to the community often fails to be accepted, since the crop lacks some characteristic favored by the farmers (e.g., taste, storage characteristics). The approach ignores local preference and the farmers' "place-based knowledge." But some organizations, through diligent experimentation and thought, are beginning to forge links with local farmers. Centro Internacional de Mejoramiento de Maiz y Trigo (CIMMYT), one of the CGIAR institutions, provides an example (Cash et al. 2003; Humphries et al. 2000). "CIMMYT has made new research findings on crop breeding useful and converted the tacit knowledge of traditional farmers into information useful to crop breeders" (Cash et al. 2003: 8088). Effectiveness is increased when information flows both ways.

Sustainable Transitions Elsewhere

There are examples of sustainable transitions taking place elsewhere in East Africa. One of the most notable is documented in a series of studies undertaken by Mary Tiffen, Michael Mortimore, and their coworkers in Machakos District, Kenya (Tiffen 1991a, 1991b, 1992, 1993; Gichuki 1991; Mortimore and Wellard 1991; Rostom and Mortimore 1991; Kaluli 1992). I need to make some general comments about Machakos District to make it clear that in many respects it differs fundamentally from Tanga Region. Machakos Town, gateway to Machakos District, is only 65 km from Nairobi, the national capitol. Dianne Rocheleau (pers. comm., 7–10 March 2004) called it "hyper-connected." Its central location means that it has long been involved in national politics. Historically, Akamba formed the major source for members of the King's African Rifles (Parsons 1999). Akamba men fought in North Africa and the Asian theater in World War II. Those experiences showed them that they were the equals of white men when it came to fighting, repairing machinery, driving vehicles, etc. This "de-mystified the white colonists and their abilities to manage these technologies" (Rocheleau, pers. comm., 7–10 March 2004). Machakos District was a popular field for Christian missions. In the period 1940–1980 it was a national priority area for food relief and development. Many nongovernmental organizations (NGOs) have long been active in the district.

In the years between the end of the Mau Mau "emergency" and independence (1955–1963), the British colonial government in Kenya adopted new policies regarding farmers living on African reserves. The idea was to register land to individuals in economically viable units, to intensify agriculture, and to create wealth for the farmers. In his report *A Plan to Intensify the Development of African Agriculture in Kenya*, Swynnerton stated: "Former government policy will be reversed and able, energetic or rich Africans will be able to acquire more land and bad or poor farmers less, creating a landed and a landless class. This is a normal step in the evolution of a country" (1955: 10). Whether the process is "normal" or simply callous can be debated. In any event, the process of registration took place in Machakos District, and from it emerged families with freehold title to land, as well as many with no title or access to land. Swynnerton's prediction on creating a landed and a landless class came true.

Improvements on registered land were funded by military remittance income as well as income from Akamba who worked in urban areas or performed off-farm wage labor. One result was an increase in the number of landless Akamba and massive migration within Ukambani and elsewhere to settlement schemes (Makueni) and to various Crown Land areas the British had previously closed to settlement (Yatta). This migration still continues and is a structural by-product of the "sustainable transition" that has occurred elsewhere in the district (Rocheleau, pers. comm., 7–10 March 2004). Like the inhabitants, the land itself was affected. There certainly was need to do something about the state of the environment in Machakos District (Maher 1937; Dunne 1979; EcoSystems Ltd. 1986).

I did research in Machakos and Kitui Districts in 1961–1962 (and again in 1978–1979) and saw firsthand the problems of soil erosion (figs. 8.3 and 8.4). In Ukambani, the land of the Akamba, the people had long experienced constraints on their livelihood because of massive land alienations and limitations on their mobility (Rocheleau, Benjamin, and Diang'a 1995; Porter and Sheppard 1998: 344). Prior to independence there was great resistance to British policy, which required the building of terraces and control of livestock (Mackenzie 1998). The resistance to livestock control already had a long history. In 1938, the government backed down when a protesting crowd of 2,000 angry Akamba men, women, and children confronted the governor, who then cancelled a plan of forced destocking (Porter 1979: 47). Resistance to forced building of terraces was also

FIG. 8.3. Top soil eroded from an unterraced hillside covering a road in Ngelani, Machakos District, Kenya. This erosion resulted from a single storm. Photo taken 14 November 1961. The Akamba resisted terracing prior to independence, which came in 1963, but embraced terracing after that. I saw this hill again in 1989, and it was fully terraced.

strong and reminiscent of that seen in Tanganyika prior to independence, although women did not go on strike (Bahemuka and Tiffen 1992: 15). Ukambani was famous for its environmental degradation, erosion, and deposition in riverbeds (fig. 8.5). When Arthur Creech Jones, undersecretary of state for the colonies, visited Ukambani in 1946, the *East African Standard* reported:

> In a tour through the Kamba Reserve on Monday he drove mile after mile through hillsides and plain swept bare in many places to the solid rock, through areas where there was hardly a vestige of grass, through acre after acre of dead wilted maize.
>
> "It was a depressing sight," he commented to Press representatives. And when he made that comment he had seen by no means the worst. . . .
>
> The route followed led through increasingly depressing country—even the top of Mbooni hill at an altitude of 6,200 feet was parched and worn out. . . .
>
> The worst eroded areas were undoubtedly those passed after the descent from Mbooni, where a fair comparison can be made with parts of British Somaliland—sand rivers, where water can be obtained by digging, lying between

FIG. 8.4. Soil erosion on a hillside in Ngelani, Machakos District, Kenya, just north of Ngelani market. The erosion resulted from a single storm. Photo taken 14 November 1961.

FIG. 8.5. Mr. Nzuki, retired army warrant officer, shows where his father's farm was and the width of a stream he used to jump across when he was a small boy in about 1945. Photo taken in Katse, Kitui District, Kenya, 6 April 1979.

bare red hills. In some places even flat plains were so worn out that there was no grass, and sand rather than top-soil. (Quoted in *Machakos District Gazetteer* 1958: 22–23)

Yet after independence (1963) things began to change in Ukambani. First, many men were absent from the area, working elsewhere in the country. They may have returned from time to time to help in farming and sent money to their families, but women assumed a greater decision-making role in the farm and the household. The *mwethya*, the traditional Akamba work party, came to be institutionalized after independence. The traditional *mwethya* had been ad hoc, a short-lived gathering of people who carried out a job of work (such as weeding) in exchange for a party with some beer and food at the end. Under women's leadership, the *mwethya* came to be an ongoing organization with long-term goals and a recognized leadership. They were registered with the Community Development Department so that they could operate bank accounts. Still another self-help institution was called *harambee*, which in Kiswahili means "let's pull together." These were festive gatherings organized periodically, with dancing, drumming, and singing, in which people urged one another (and visiting politicians and dignitaries) to make donations in cash or in kind (eggs, chickens, goats) to further some project. They were strongly supported by Presidents Kenyatta and Moi, and they were instrumental in raising large sums of money for local development.

After the corrupt 1988 elections, many Akamba were "upset at the government and questioning its legitimacy" (Rocheleau, pers. comm., 7–10 March 2004). Men and women would no longer turn out for *harambee* occasions or chief's *barazas* (meetings). Women expressed their disfavor over the way their *mwethya* groups were co-opted for political ends by resigning from them—by disestablishing the *mwethya*. They continued to work together in groups of two to five people, so they were still linked and organized.

The most notable development in Machakos District between 1963 and 1990 was a return to terrace construction, much of it done by women as part of *mwethya* work parties. Considerable use was made of the *fanya juu* (literally, "do it upward," in which soil is thrown upslope until one has formed a bund along the contour that will interrupt the flow of rainwater). In due course the sloping land fills in behind the bund and the terrace becomes more level (United Nations Environmental Program 2000: 1). Rostom and Mortimore documented that an astonishing 100 percent of all agricultural land in four locations was terraced between 1948 and 1978 (Rostom and Mortimore 1991: 8). This was generally true throughout Ukambani, although the areas of lower agricultural potential were not terraced so comprehensively. Nonetheless, terraces continued to be favored in more marginal areas because they are such excellent water-harvesting devices. Maize yields are higher and more reliable when grown on such terraces.

One result of land consolidation and registration was a decline in polygamy and the emergence of nuclear families as the norm and a decline in the importance of clan membership (Bahemuka and Tiffen 1992: 7). De facto female-headed households were also important in Machakos District. The 1979 census showed that 103,185 Akamba men lived in Nairobi District, a preponderance of whom came from Machakos District (Tiffen 1991a: 25).

Over the years there was a marked increase in technical staff promoting agriculture. In 1950 there was 1 agricultural officer and 80 assorted technical officers, technical assistants, and junior technical assistants. By 1988 there were 19 agricultural officers and 143 other technical officers (Tiffen 1991b: 27). Technological change and innovation in Ukambani over the decades was considerable and it came from many sources, but most notably from Akamba farmers themselves. It was they who figured out ways to use ox-plow teams on terraces, and they experimented energetically with maize varieties (Rostom and Mortimore 1991: 43; Mortimore and Tiffen 1994). In the period 1930–1990 great increases were made in maize growing, in the production of fruit and horticultural crops, ox-plow tillage, the use of farmyard manure, composting, mulching in agriculture, adoption of grade and cross-bred cattle, and stall feeding. All these developments have led to greater flexibility or ability to choose among adaptive measures for Akamba farmers.

By 1987 enrollment in primary schools was over 390,000 students, which represented over 28 percent of the total population (Tiffen 1991b: 82; Bahemuka and Tiffen 1992: 25). In 1982 literacy rates were 68 percent among men and 42 percent among women. Also important were Village (or Youth) Polytechnics, low-cost training centers designed especially for school dropouts. Adult education was quite active, especially the women's literacy effort. In 1987 there were 80 active cooperative societies. Most rural families belonged to self-help groups and were church members (Bahemuka and Tiffen 1992: 45). In 1988 there were 16 NGOs active in Machakos District. They were involved in soil and water conservation, water supply and sanitation, community health, tree planting and fuel wood and fruit tree provision, agriculture and livestock development, and supplying improved woodstoves (Kaluli 1992). The net result of all this activity was that a large cadre of educated leaders emerged throughout urban and rural Akamba society. As Bahemuka and Tiffen observed: "By comparison with the colonial period there is more communication and confidence between governed and government, and most villages have both more social amenities and more knowledge with which to deal with their difficult environment" (1992: 43). The efforts of agricultural extension staff have been judged less successful (Bahemuka and Tiffen 1992: 31).

People felt more in control of their lives and their surroundings. More people were living on the same amount of land, yet conservation measures seemed to have checked erosion and soil loss. There was not a marked increase in agricultural productivity (i.e., yields), but some crops did increase in prominence, and these tended to be for external markets: tomatoes for urban markets and export by air to Yemen, and Asian vegetables (eggplants, chilies, cilantro, sorrel, spring onions, various greens, etc.) for the large Asian market in Nairobi and Mombasa (Tiffen 1991b: 34).

Not everyone agrees with the positive picture of accomplishment presented by Tiffen and her coworkers. Murton has faulted Tiffen and her coworkers for their reliance on aggregate, district-level data. He did a longitudinal study of 180 households in a village in Mbooni Location and found that those who had succeeded in their farming were largely those with access to nonfarm income (usually urban derived) that they could invest on-farm, thus creating a virtuous cycle (Murton 1999: 37–38). Those without off-farm income experienced declining yields, declining soil fertility, and lower labor productivity (Murton 1999: 44). These people were undergoing an agricultural involution, à la Clifford Geertz, while neighboring land-rich farmers were accomplishing a Boserupian intensification.

Although a greater population now lives in a landscape in which conservation measures appear to have checked erosion and soil loss, other developments are cause for concern. There is now a large landless class and virtually no *weu*, or common land, in Machakos District. Women in particular have difficulties in their livelihood, since they usually do not hold title to land and may not be authorized to make decisions. The standards of terrace maintenance, soil conservation measures, and watershed management have declined (Rocheleau 1991; Rocheleau, Benjamin, and Diang'a 1995: 218; Murton 1999: 37; Batterbury and Forsyth 1999: 27–28).

It is not clear that an enduring sustainability transition has taken place in Machakos District. If such a transition has taken place, it has left a lot of people out of the process. The number of landless and land-poor people and people in poverty is large. Migration to marginal areas continues at a high pace, but it is very hard to make a go of it in these areas. As one man living in an agriculturally marginal area told Dianne Rocheleau: "The children can grow up here but they can't stay here" (Rocheleau, pers. comm., 7–10 March 2004). Economic differentiation has increased greatly. All of the preceding results are treated as externalities in the characterization of the successful sustainable transition in Machakos District. The prospects for many in the next generation are not good.

A Sustainable Transition for Tanga Region?

Having reviewed an instance of considerable, sustainable, development in Machakos District, let us explore the matter as it applies to Tanga Region. It would seem that conditions that are enabling for development include the following: a population that is literate, educated, able to organize into local self-help groups, unconstrained by bureaucratic red tape, and able to gain access to funds and technical advice. Land registration may be an important ingredient. How do the people of Tanga Region measure up, compared with the folk in Ukambani? What of the institutional support from government and NGOs?

Like Machakos District, Tanga Region has experienced erosion and declines in soil fertility, though not at such spectacular levels. Although beliefs in the causes of soil erosion and land degradation in Africa are so ingrained as to become myths or "environmental orthodoxies" (Forsyth 2003: 50), there is little disagreement that erosion and soil depletion in Tanga Region are physical facts (Ngware 1982; Ngatunga 1981; Scheinman 1986; Conyers et al. 1970: 113; National Soil Service 1989b: 3–7; Kaihura 1991: 8). Large-scale logging operations have led to considerable erosion and increased runoff; cardamom cultivation reduces soil fertility and makes it necessary to open new fields after 8–12 years (Hamilton and Bensted-Smith 1989: 32–33; National Soil Service 1989b: 6). In the lowlands, thousands of hectares on sisal estates have been cropped for long periods, some as much as a hundred years, leading to decline in soil fertility (Hartemink 1995). Decades of land clearing have transformed the once-forested landscape in higher areas (fig. 3.7).

For the most part, customary land tenure continues to prevail in rural areas in Tanga Region. Land parcels that belong to farmers are frequently widely scattered. The philosophy underlying *ujamaa vijijini* was antithetical to individual freehold of land.[3] Nothing comparable to the registration of land and land consolidation that took place in Machakos District has occurred.

In 2002 the census showed that there were 80 men for every 100 women in the age cohort 20–39 in Tanga Region (United Republic of Tanzania 2003). Although some of this may have been due to death from AIDS, much of it had to do with men working outside the region. There has been a decline in polygamy among the Sambaa in the Usambara Mountains and this decline may be more widespread (Feierman 1993: 126). Like the Akamba, the people of Tanga Region have the equivalent of the *mwethya,* or work party. It is called the *kibarua* and is frequently used, particularly by wealthy farmers. I saw no evidence that the *kibarua* had been institutionalized as it had become in Machakos District.

In Tanga Region in 1988, 12 percent of the total population, or 37 percent of youngsters (ages 5–18) attended school, and only 118 students were enrolled in postsecondary educational programs. Boys were more than twice as likely as girls to be enrolled beyond high school. About 52 percent of men were said to be literate, and 41 percent of women (Bureau of Statistics—Takwimu 1992: 70, 98, and 57). The sorry state of schools is suggested in figs. 5.6 and 5.7. Several weeks after I took those pictures, we visited the Amboni Sisal Estate, long run by the Swiss. They had a school for the children of the small expatriate staff. We visited the school, a spacious two-story house with four large classrooms. "They were just installing the computer which has CD-ROM capabilities. On Jim's desk were sitting several laser disks, each holding the equivalent of the works of an encyclopaedia—games, maps, nature, biology, mathematics disks" (Porter and Porter 1993: 88). From 1986 onward, government funds to support education were cut sharply. As a cost-sharing measure, school fees were reinstated (Doro 1995: B355). Tanzanian Library Services subsequently suspended its service in Tanga Region because of a lack of funds. Thus, there were no books or newspapers available through libraries (Kiondo 1999).

It was a common experience of African countries, as they emerged from their status as colonies, to develop governments that looked outward rather than inward for resources for development. The elites that assumed the responsibilities of governance, in many cases, became rent-seekers, more linked to foreign donor governments and foundations and to foreign

NGOs from which development funds could be obtained than they were to the general populace. Governments became "suspended," that is, disarticulated from the people, separated from the "bed-rock social forces of African life" (Samatar and Samatar 1987: 670). Dunstan Spencer has stated that the key to small farmer development, even survival, in the long run will be to "[p]rovide national governments with the means to invest in people-centered development, i.e. in education, health and rural infrastructure (roads, irrigation systems, markets, post harvest machinery, etc.)," and to "reverse the disgraceful downward trend in support to government research and extension systems" (2001: 2). Tanzania's government has not been a suspended state and it made great efforts to prevent a "rent-seeking" elite from emerging in the early years of independence. In 1971 TANU published a new leadership code (*mwongozo*) that forbade civil servants and party members from having outside income (Tripp 1997: 6; de Souza and Porter 1974: 66). Nonetheless, connecting government with the rural population had mixed results—excellent with respect to adult literacy, disappointing with respect to agricultural development. I recall President Nyerere in the 1960s at the time of NASA's Apollo Program saying: "[W]hile other people can aim at reaching the moon and while in future we might aim at reaching the moon, our present plans must be directed at reaching the villages" (Nyerere 1966: 131). Yet they have had a difficult time of it.

Given the fact that national governments find themselves unable to fund many development efforts, they turn to NGOs for assistance (National Steering Committee for NGO Policy 2000). Tanzania Non-governmental Organizations (TANGO), founded in 1988 by 22 NGOs, listed over 560 organizations that were working in the country in 2004. A number of these work in Tanga Region and are active in a wide variety of programs, for example, forestry, traditional medicine, family health, distribution of improved seeds, marine science, mosquito control, aid to street children, training in use of agricultural machinery, village development, irrigation, soil erosion, and agroforestry.

As to the question of a large educated leadership in Tanga Region, employment figures can suggest the size of this group. Table 8.1 shows that only 10 percent of the people in Tanga Region were employed in pursuits other than farming or agricultural labor, thus providing a vivid contrast with Machakos District.

Some idea of the degree to which local women in Tanga Region are connected with extension agents of various kinds and with useful information

TABLE 8.1. Employment Status of People Aged 10 Years
and Over, Tanga Region, 1988

Employment class	Number
Legislators, administrators, managers	2,117
Professional, technical, teachers	19,906
Clerks	5,548
Service and sales persons	17,781
Cultivators	463,109
Mixed farming	25,243
Agricultural workers	2,613
Craftsmen and machine operators	15,062
Small-scale traders and laborers	23,050
Other workers	5,392
Not employed[a]	282,559
Total	862,380

Source: Bureau of Statistics—Takwimu 1992: 161.
[a]Of which 152,132, or 54 percent, were under the age of 15.

is found in Elizabeth Kiondo's writings (1999). She interviewed 773 rural women and 64 information providers in three districts in Tanga Region. Those interviewed were women mostly aged 26–35; of these 31 percent were illiterate, 44 percent earned less than $100 a year, and 30 percent belonged to a women's group. According to Kiondo: "The critical role of women in development has not been part of mainstream development thinking and strategies in Africa." Women do not get the information that men do, and "training is mainly imparted to male members of rural societies." Women get their information orally, usually from family, friends, and neighbors, as they fetch water, cut firewood, go to markets, or attend weddings and funerals. They seldom see experts of any sort; 70 percent had not been visited by an agricultural agent, and 95 percent had never seen a cooperative worker. Health information is more readily available for women (reaching 87 percent of them) than information about agriculture (28 percent). About 78 percent of information providers were men. Women's access to agroforestry and soils information was very low— 12 percent in Muheza District and 0.5 percent in Korogwe District. Further, Kiondo stated that the "1988 National Population Census in Tanzania indicates that women head 42.2% of rural households." Taking the number of single, divorced, and widowed women aged 20 and over, I calculate the figure to be about 51 percent of the total number of households (Bureau of Statistics—Takwimu 1992: 47, 231).

Probably the weakest part of the knowledge base in this study concerns economic activity that affects farmers but is organized by others, mainly outsiders, at the time of harvest. In Chanika (Handeni's nearby commercial center) lived 30 trader families (Hindu and Ismaili), who ran shops, bought crops from farmers, transported them to other markets, and sold goods to the local people (Sadleir 1999: 115). In other settlements throughout Tanga Region one can find Asian or Arab traders and shopkeepers. Many entrepreneurial young men from Lushoto, Tanga Town, Moshi, Arusha, Dar es Salaam, Mombasa, Nairobi, and other places visit different parts of Tanga Region on a regular or seasonal basis. For example, consider the development of the trade in oranges. Oranges were introduced in 1937 by missionaries, but after 1955 they were spread through the efforts of the Mlingano research station, which provided improved budding material (Regional Planning Office 1983: 18). At first the oranges were sold in Mombasa. After 1976, with the collapse of the East African Community, the markets shifted to Dar es Salaam, Arusha, and Mbeya. Lorry owners come from far away and help harvest and buy up the oranges. Some local farmers organize their own sales, rent a lorry, and take the oranges to Dar es Salaam, Moshi, or other cities. The local trade in oranges is done by the local people. However, much of the wholesale trade is done by Sambaa.

These "visitors" are active as traders, transporters, and middlemen in every sort of trade, buying in bulk at harvest time and moving goods to market, frequently black-market destinations. Others show up at weekly or periodic markets to sell used clothing, pots and pans, and other hardware. I know that these entrepreneurs are an important part of the economic system in Tanga Region, both enabling and disabling in their effects on local farmers. Since our questionnaire was not designed to get the historical details of marketing of produce, and since we did not interview these entrepreneurs, I can only cite what other observers have said about this "external" set of actors in the economic life of Tanga Region.

If we run through the roster of conditions that changed in Machakos District, Kenya, over the years, we find the people of Tanga Region at an early stage of such changes. The roster includes registration of land, emergence of nuclear families, institution of the *mwethya* as a permanent self-help organization with leadership and long-term goals, and fundraising through an institution like *harambee*. It also includes extensive soil and water conservation work, increase in numbers of technical staff, help from NGOs, technological innovation in farming and stock rearing, and increased rates of education. If developments such as those that happened in Machakos District occur throughout Tanga Region, they will benefit many, but they will also likely lead to increased social and economic differentiation and increased poverty and disempowerment among women and poor families. In any event, Tanga Region could never have the same history that Machakos District experienced.

Two of the anonymous reviewers of the manuscript of this book took me to task for presenting a "generic Marxist statement" regarding some of the outcomes (the unaccounted-for externalities) of the sustainable transition in Machakos District—increased poverty, landlessness, and migration to marginal areas. They argued that in all economic development there are winners and losers. They wanted to know my opinion about the Machakos story and what I recommended for Tanga Region. One reviewer commented that an analysis of contemporary American agriculture would conclude that family corporate farms do well, while some rural communities disintegrate. Structural and social adjustments occur constantly in labor and economic sectors, thus creating winners and losers. This reviewer noted that it is preferable for half the population to achieve higher living standards, while at the same time ensuring the stability and productivity of the environment on which those standards are based, than it is to make marginal improvements for everyone.

If Tanga can undergo a sustainable transition only if it incurs structural-social adjustments in which a significant number of farmers are losers and must leave their farms, we face a Rawlsian dilemma (Rawls 1971). To what extent should one accede to increases in economic differentiation if the total sum of economic good is greater? I noted earlier that there have always been rich, powerful families and poor families in rural Tanga. Further, the current government of Tanzania has apparently abandoned its earlier goal of fostering a society in which economic differentiation is minimized (*mwongozo*). Conditions are present for large-scale structural and social change.

The Task Ahead

I will let a definitional statement by Nancy Peluso and Michael Watts set the stage.

> Our starting point is the entitlements by which differentiated individuals, households, and communities possess or gain access to resources within a structured

political economy. It grants priority to how these entitlements are distributed, reproduced, and fought over in the course of shaping, and being shaped by, patterns of accumulation. All forms of political economy have as their foundation the transformation of nature in social, historical, and culturally informed ways—what Marx broadly construed as the labor process. We start by seeking to understand the changing contexts of nature transformation, who performs the labor, who bears the burdens, and how benefits are claimed, distributed, and contested. (2001: 5)

Tim Forsyth places a great emphasis on *framing* as an element of research and policy formulation: "The local perception or evaluation of environmental changes may be referred to as 'framings.' This term refers to the principles and assumptions underlying political debates and actions" (2003: 77). Other observers of the environment—Western-trained foresters, hydrologists, agronomists, economists—frame the environment in different ways, sometimes in ways called "environmental imaginaries" (Peet and Watts 1996; Zimmerer and Bassett 2003). Such terms as "desertification" or "deforestation" or "shifting cultivation" embody and simplify a host of presumed attributes, usually negative. They are taken to represent established facts (reality), when rather they are particular framings of an observed environment. It is important that local views about environmental changes and their consequences be considered, that they be framed.

In this research, I made every effort to frame the approach to drought hazards in Tanga Region so that data and ideas emerged from what people told me. They defined and elaborated on the nature of the risks they face and they defined what might be termed "normal risk" and "crisis risk" (Forsyth 2003: 191; Sperling and Langley 2002). It has been my purpose to examine crop-water availability and its consequences for farmers in 18 villages in a variously endowed transect across Tanga Region. Crop nutrition—fertilizers, soil pH, cation exchange capacity, and such matters—though important, was not the focus of this study. What do farmers do about *water* for their crops? By focusing on decisions a farmer takes in growing a crop, I was able to show that in nearly every case, farmers made good decisions. These good decisions were reflected in the choice of crops to grow (green grams vs. beans, cassava vs. maize, etc.), which crops to interplant, where to plant crops (*mabonde*, lower slopes, etc.), which season to use, when to plant the crop, and what planting density to use. A series of simulations explored each of these questions. In most instances,

I found that local practice either *matched* the recommendations of Western agronomists or *beat* the recommendations of Western agronomists or *maximized* some goal of local farmers, such as yield reliability or coping with labor bottlenecks, that may have been overlooked by Western agronomists. Is there a way for Western agronomy to take local insights and locally proved performance into account? Is there some way women and men farmers could be extension agents to the research stations? The United Nations Inter-academy Council's report *Realizing the Promise and Potential of African Agriculture* refers frequently to the need to incorporate local knowledge and local farmers into agricultural research and development (Inter-academy Council 2004: xxii–xxiii, 142, 164, 212, 228). I have no answers to these questions, but I would like to see these things tried. Although there is some transfer of knowledge between the two scientific cultures I have discussed in this study, there could be more. Would it be possible to commingle expert knowledge, despite its commonly encountered paradigm of ecological fragility, with "alternative knowledge" and local adaptive process (Forsyth 2003: 29)?

Examples of successful cooperation between staff of research institutions and local people are embodied in the growing adoption of PAR—participatory action research, inspired by the methods of Paulo Freire (1970; Chambers 1997: 106ff.; Sperling in Smith et al. 2004).[4] Chambers notes the great abilities of local people: "Like others, I had been astonished by the analytical abilities of poor people. Whether literate or not, whether children, women or men, they showed that they could map, list, rank, score and diagram often better than professionals" (1997: xvii). He summarizes the contribution of PAR in terms of the normative ideas it generates. The most important of these are "(1) that professionals should reflect critically on their concepts, values behaviour and methods [their framings]; (2) that they should learn through engagement and committed action; (3) that they have roles as convenors, catalysts and facilitators; (4) that the weak and marginalized can and should be empowered; and (5) that poor people can and should do much of their own investigation, analysis and planning" (1997: 108).

The possibility exists to improve Tanga's agriculture and the well-being of its people. In Tanga Region, those who *study* and those who *tend* growing plants (life's delicate children) could benefit by working together and learning from one another. Then, perhaps, the promise in the name of one of this study's 18 villages (Maranzara, which translates as "end hunger") will be realized.

Agrometeorological Modeling

NEWMODEL, a suite of models of water-energy-soil-crop relations, is available online at www.geog.umn.edu/, under faculty, Porter. At that site one is provided with an explanation of how the program works and the biophysical knowledge upon which it is based. One can download the operating program itself, along with precipitation, evaporation, and crop data for a number of stations in East Africa and readme files that explain the use of the various programs. The documentation is divided into three parts. Part I describes data requirements and Part II sketches out the biophysical assumptions of the program. These parts are reproduced here. Part III, Program Simulation, explains the sequence of steps the program NEWMODEL.EXE goes through in its simulation. It can be read online at the above-noted Web site.

Part I. Data Requirements

NEWMODEL is a menu-driven program that estimates the consequences of moisture stress on subsequent crop yields. It is flexible in permitting the user to specify the nature of the atmospheric and soil environment in which the crop grows.

NEWMODEL was written by John D. Corbett (Corbett 1990). The program was made "user friendly" by Gregory M. Flay, Department of

Geography, University of Minnesota, in 1991–1992. NEWMODEL was developed under the general supervision of Philip W. Porter, Department of Geography, University of Minnesota, and is intended particularly for use in study of agriculture in semiarid and subhumid lands in East Africa.

The model consists of a group of related programs (in addition to NEWMODEL) which enable the user to modify data and to explore rainfall records in a variety of ways. The programs, identified by extension *.EXE, are

> NEWMODEL.EXE
> NEWVIEW.EXE
> PRECIP2.EXE
> READMNTH.EXE
> DISTRIB.EXE
> BIN2ASC.EXE

Each program is described in this appendix but is also documented internally in the program and can be used without this appendix.

NEWMODEL.EXE

NEWMODEL requires a number of files in its analysis. These are identified by extensions, which makes searching and management easier.

PRECIPITATION STATION FILE.   The file 9538011.PRE can serve as an example.

Station no.	Name	Lat.	Long.	Elev. (m)
9538011	Mlingano Agri. Inst.	5′09°S	38′54°E	188

The identification of stations in NEWMODEL differs slightly from that used by the Tanzania Directorate of Meteorology. For example, 9538011 would be written 095.3811 by the Directorate of Meteorology. As used here, 95 refers to a square, 1° of latitude by 1° of longitude, 95° of latitude south of the North Pole; 38 refers to 38° of longitude east of Greenwich meridian; and 011 indicates that Mlingano was the 11th station to be established in the 1° square. The station, in other words, is found in a 1° square bounded north and south by latitude 5° south and 6° south and bounded on the east and west by longitude 38° east and 39° east of Greenwich.

Another difference is that each station file is identified with an extension: .PRE, standing for precipitation.

EVAPORATION STATION FILE. These files contain daily evaporation values. For example, the file MLINGANO.EVP, for Mlingano, is based on daily readings of pan evaporation over a period of 3 years (1978–1980) and smoothed with a 21-day running mean. The file is based on US Weather Bureau Class A open pan evaporation data (fig. 3.5).

CROP COEFFICIENT FILE. This file contains daily E_t/E_o values for a variety of crops, mostly annual cereals and legumes grown in East Africa. For example, the file HM150.CRP describes the coefficients for a hybrid maize that takes 150 days to mature. E_t/E_o stands for the ratio of crop transpiration to that of an open water surface (E_o). Values are low at the beginning of a crop cycle and may rise to 1.0 or more at the time of flowering and maximum leaf area.

In addition to files for precipitation, evaporation, and crop coefficients, one needs to know other facts (or be able to specify assumptions) about the crop and soil. These are (1) planting-date parameters, (2) agronomical parameters, (3) full cover coefficient, (4) plant spacing, (5) daily yield reduction curve, (6) SCS curve number, (7) soil characteristics, and (8) any assumption regarding delay in planting. These are described in detail in part III (see Web site).

Part II. Biophysical Reasoning

To sketch the reasoning behind the water-energy-soil-crop model, imagine two biographies: first, that of a raindrop in its journey through the hydrologic cycle and, second, that of a plant, from its germination to its physiological maturity and subsequent harvest. The dominant metaphor in these biographies is that of *branching,* that is, taking this road and not that road.

The raindrop falls from a cloud. It lands on a plant or on the ground, perhaps bare soil. If it lands on a plant, it eventually evaporates. If it lands on soil, it may move in any of five different ways: it evaporates, it becomes bound to soil particles, it runs off on the surface, it migrates through the soil column (past the plant's rooting zone) and leaves the system as drainage, or it gets taken up by the roots of the plant. If it is taken up by the plant, it is used either in photosynthesis, adding biomass to the plant, or it is used in transpiration, to help regulate the plant's temperature. Which of these events occurs in the life of the raindrop depends on the state of the atmosphere and, especially, the soil at the time the drop hits the ground.

With a fully saturated soil, the raindrop will likely be part of runoff. If the soil is very dry, there is about a 50 percent chance that it will evaporate on the first day after the rainstorm. What happens next depends on the atmosphere (wind, relative humidity, etc.) and the soil (texture, amount of carbon, how deep it is, and the soil moisture status). Soil moisture is determined by a combination of relations among bulk density, field capacity, and permanent wilting point.

Moisture loss from the soil (by evaporation) is modeled independently from moisture loss from the plant (by transpiration). On a bare, unvegetated soil, moisture loss by evaporation will be dominant. Over the course of about 3 weeks, a soil that starts out wet will dry out, so that by the end of the period, in most instances, a vapor barrier is formed, preventing moisture from moving toward the surface by capillary action and then vaporizing into the atmosphere. Thus, over a period of 20 days, E_s/E_o (E_s being evaporation from the soil) will decline exponentially from 1.0 to 0.01. If a rainstorm occurs during the 20 days, the system "resets" to day 1. As plants become bigger and begin to shade the soil, the potential rate of moisture loss from the bare soil is reduced. With full vegetative cover, there is almost no evaporation from the soil.

These are assuredly complex matters, but they have been studied by meteorologists, soil scientists, and agronomists and have been modeled. All these processes have been described numerically.

As for the biography of the plant, one wants to know about the status of the soil at the time the seed is planted. Next, one wants to know what happens to it day by day through its life course, with particular attention to its well-being at critical, potentially stressful moments, such as flowering. One needs to know the context in which it is maturing: plant spacing and soil characteristics and depth, all of which combine to determine the soil's available water capacity (AWC), as well as the total amount of water available to the plant on a given day.

When its starts out, the plant has modest needs for water; but although it does not need much (compared with when it has reached its full vegetative growth), a shortfall of moisture at germination can be very stressful and jeopardize its life path. When it reaches its full development (as measured by its LAI, or leaf area index), the plant flowers. Moisture stress at this time can have devastating effects on subsequent yield. In maize, for example, the male pollen dusts the silks that lead to the sites where the kernels will grow. If, because of moisture stress, fertilization fails to occur as it should, the result may be an ear of maize with only a few fully developed kernels and many that are undeveloped. On the other hand, if fertilization is accomplished fully, but the plant experiences a moisture shortfall during the grain-filling (or seed-filling) period, the kernels may be small. As the season progresses, more and more crop energy goes toward the seed. Photosynthetic material is translocated from leaves and stems to the seed. (There are plants, especially grasses, that translocate photosynthetic material to the root systems toward the end of their life cycle.) The plant requires less water than it did earlier in the season, and still less during the ripening period, as it approaches physiological maturity.

These two biographies give some idea of the elements that have been specified in a quantitative manner in NEWMODEL. When the model is run, the program keeps track day by day of the status of all these variables. It is even possible to print out a "diary" of these variables, since a running account is kept of all changes.

Other supplementary programs (NEWVIEW.EXE and PRECIP2. EXE for graphic display and READMNTH.EXE, DISTRIB.EXE, and BIN2ASC.EXE for data preparation, manipulation, and analysis) are explained fully on the Web site. NEWVIEW.EXE generates water-balance diagrams using random-access files created by NEWMODEL and precipitation and evaporation station files (fig. 6.3). PRECIP2.EXE creates a graph of average monthly rainfall for a given precipitation station. The graph assigns one dot per recorded precipitation value and draws a trend line representing the average precipitation for each month. All data are in millimeters (see fig. 1.1 and figures in chapter 5). READMNTH.EXE calculates total rainfall occurring in each month using a daily record of precipitation values and evaluates whether a given month is a dry or a wet month, depending on a definition of "dry month" provided by the user. DISTRIB.EXE produces graphs from *.TBL files generated in the crop yield simulation program NEWMODEL (fig. 6.2). BIN2ASC.EXE converts binary precipitation data into ASCII text files. Binary files take less disk storage space.

In this abbreviated appendix I have provided only a basic introduction to NEWMODEL, including a sketch of its biophysical underpinnings.

Agroclimatological Survey
of Tanga Region

Life History, Crops Grown and Changes, Environmental Assessment, and Coping Strategies

1. Date: _____ 2. Time: _____ 3. Interview No.: _____
4. Interviewer: _____ 5. Recorder: _____
6. District: _____ 7. Division: _____ 8. Ward No.: _____
9. Ward name: _____ 10. Village name: _____
11. *Balozi*'s name: _____

In 1972, some students from the University of Dar es Salaam and I inter-viewed ( ) people in ( ). At the time we could not talk with everyone in the division, and we selected those to be interviewed by chance. Now I have returned, 21 years later, to talk with family members of those we interviewed then and, in some cases, the people themselves whom we met.

As with the earlier survey, in this questionnaire, we want to find out about farming and stock keeping here and how they are different now than they were in 1972. We also are interested in how you use the land now, in comparison with 21 years ago. The information about how you make a living, particularly about how you grow your crops, will be compared with detailed information about the climate of this area, especially the rainfall and the variability and reliability of the rainfall. We hope that your answers will enable us to recommend farming practices and crop varieties which will improve the productivity of agriculture and raise the income of the people of the region.

I am from the University of Minnesota, in the USA, and while in Tanzania I am affiliated with the Institute of Resource Assessment, University of Dar es Salaam. I will very much appreciate your cooperation. If you do not wish to answer any question I ask, that is fine. If at any time you wish to stop the interview, you may. You have my assurance that you will remain anonymous in the book I write about Tanga, and the answers you give will never be connected with you personally.

12. Respondent's name: _____

13. Relation to individual interviewed in 1972: _____

Interview No.: _____

For office use:

14. Enumeration Area No.: _____   15. Pop./sq. km: _____

Life History of Respondent

16. Where were you born? Village: _____   Division: _____
District (if needed): _____

17. How old are you? _____

18. How many years have you lived in this place? _____

19. Before coming here to live, where did you live? (Ask only if applicable.)

Years	Village	Division	District
_____	_____	_____	_____
_____	_____	_____	_____
_____	_____	_____	_____

20. What big towns have you traveled to? _____

| _____ | _____ | _____ | _____ |
| _____ | _____ | _____ | _____ |

21. Have you ever worked for wages? _____

22. If yes, what was the total time you were working for wages?

_____

23. Where did you have this/these job(s)? _____

_____

24. Which of these jobs did you have the longest? _____

25. What is your occupation now? _____

26. How many years of formal schooling have you had? _____

    Questions 27 and 28 removed during field trial.

29. Have you been a member of a cooperative society? Yes: _____

    No: _____ If yes, give name of society: _____

30. Are you still a member? Yes: _____ No: _____

31. Are you an officer in government or any society? Yes: _____

    No: _____

32. If yes, name of position: _____

Family Characteristics

We want to know about the members of your family and the amount of labor available to you to help in your fields.

33. Are you single_____, married_____, divorced_____, or widowed_____?

    Please give information on your wife (or wives) and living children. (To interviewer, list first wife and her children, then second wife and her children, etc. We don't need to know their names, but it will help in asking questions about each person to have the names. If you are interviewing a woman who is head of household, so indicate.)

34.

Name	Relationship	Sex	Age	No. years education completed*	Presently** living here	Does this person work on farm?
				Y	Y N	Y N
				Y	Y N	Y N
				Y	Y N	Y N
				Y	Y N	Y N
				Y	Y N	Y N
				Y	Y N	Y N
				Y	Y N	Y N

(Use back of questionnaire if necessary.)

 * If still attending school, circle the Y on table.

 ** If not present, query 35.

35. Does he (she) return to help with the farm work? _____
    If yes, specify season: _____
36. Is there anyone else living here? Y _____   N _____
    If yes, enter data in table below:

Name	Relationship	Sex	Approx. age	Does this person work for you?	Kind of work	Which months
				Y N		
				Y N		
				Y N		
				Y N		
				Y N		
				Y N		
				Y N		
				Y N		

The Farm: Land Use and Tenure

37. How many fields do you have currently in cultivation? _____
38. How many fields do you have which you are not currently
    cultivating? _____
39. Are all your fields together? Y _____   N _____
Now we would like to ask about how you obtained your land, if you don't
mind.
40. How many fields did you obtain from:
    Your father: _____ Other relatives: _____ Bought: _____
    Government: _____ Just took unused land: _____
41. If bought, when: _____
42. If you wanted to get more land for farming, to whom would you go
    talk to obtain permission? _____

43. Is there still land around here for new farms? Y _____   N _____
Questions 44 and 45 removed.
46. Are you using any other person's land? Y _____   N _____
47. What are you using this land for? _____
48. In some parts of Tanzania people have serious disputes over land. Do
    you think this is a problem in this area?
    Y _____   N _____
49. If yes, is the problem very severe _____ , severe _____ , moderate
    _____ , or slight _____ ?

Crops Presence

50. Please tell me **all** the crops you grow in the order of how important
    they are.
    _____
    _____
    _____

51. Are there some crops you interplant? Y _____   N _____
52. If yes, which crops? _____
53. Why do you interplant these crops? _____
    _____

54. How much land do you have in interplanted crops? _____   acres.
55. Are there some crops you grow in pure (unmixed) stands?
    Y _____   N _____   If yes, which ones? _____
56. Why do you grow them in pure stands? _____
    _____

57. How much land do you have in crops of pure stands? _____   acres
    Of your crops, which would you say are your three most profitable cash
crops?
58. Most profitable _____
59. Second most profitable _____
60. Third most profitable _____
    Which would you say are your most important food crops for you and
your family to eat?
61. Most important _____
62. Second most important _____

63. Third most important _____
64. Are there any other crops you grow only a little? _____
_____

65. What tree crops do you have? (If some tree crops have already been mentioned, ask: "What other tree crops do you have?" List in table immediately below.)
66. How many trees do you have?

Tree crops	No. of trees

Change

67. What are the main farming changes that have taken place during the time since 1972 (or since you came here)? _____
_____

68. What is the most important thing that has happened in your village since 1972? _____
_____

69. What is the best thing that has happened in your village since 1972?
_____

70. What is the worst thing that has happened in your village since 1972?
_____

71. How many years that were bad for farming have you had since 1972?
_____

Which years were they?

Year	Why bad?
____	_____
____	_____
____	_____
____	_____
____	_____

72. What things do you do now in your farming that you did not do ten years ago? _____
_____

73. Have you stopped doing anything in your farming that you did ten years ago? Y _____   N _____

If yes, explain: _____

_____

Crop Changes

74. Since 1972 (or since you came here) which crops have you stopped growing? _____

75. Since that time which crops have you started growing?

_____

76. Are there any crops you plant more of (a larger acreage)?
    Y _____ N _____ If yes, name them: _____

77. Are there any crops you plant less of (a smaller acreage)?
    Y _____ N _____ If yes, name them: _____

78. If a crop has increased or decreased, why?

_____

79. Have yields of any of your crops improved in the past 20 years?
    Y _____ N _____

80. If higher, (a) which crops? _____
    and (b) what has made it possible to get higher yields? _____

_____

81. Have yields of any of your crops declined in the past 20 years?
    Y _____ N _____

82. If lower, (a) which crops? _____
    and (b) what has caused the decline in yields? _____

_____

83. Compared with ten years ago are you cultivating more land,
    the same amount or less land?
    More _____ The same amount _____ Less _____

84. If more, how many acres more? _____

85. If more, what has made it possible for you to use more land?

_____

86. If less, how many acres less? _____

87. How long does it take to walk to your nearest field? _____

88. How long does it take to walk to your farthest field? _____
    (Interviewer: If the time difference is more than 15 minutes, ask:)

89. How long does it take to reach most of your fields? _____

90. How long does it take your family to walk to where you get water?

_____

91. How long does it to take your family to walk to where you cut fuel wood? _____

92. How does the period of lower rainfall in January and February affect your farming? _____

_____

_____

93. How many times since you have lived here have you had to replant your crops? _____

94. What were the problems that made it necessary for you to replant the crop? _____

95. Do you find any particular periods of the agricultural year to be times of labor bottlenecks, times when you can't get all the work done that you want to get done?

   Y _____   N _____

96. If yes, when? _____

   Sometimes there is so much work to be done and so little time to do it that people have work parties and give the people who help some food and drink after the work is finished.

97. Have you ever had work parties? Y_____   N _____

98. For which crops have you had work parties? _____

_____

99. Have you had any work parties this year? Y _____   N _____

Now for these crops that you named, for which tasks would you have a work party?

	100.1 Clearing	100.2 Digging	100.3 Planting	100.4 Weeding	100.5 Harvesting	100.6 Other
Crop 1						
2						
3						
4						

101. How many days this past year were you ill and unable to work? _____

102. What was the cause(s) of your illness? _____

_____

What other members of your family who help in the fields were ill during the past year?

103.1 Name	103.2 Approximate no. of days lost	103.3 Cause of illness

Livestock

We would like to learn about how you take care of your livestock.

104. Do you have livestock? Y _____ N _____

105. Generally, what is the largest number of cows you have giving milk during the *masika* rains? _____

106. Approximately how many cattle, sheep, goats, and donkeys do you have?

_____ cattle, _____ sheep, _____ goats, _____ donkeys

107. Do you have more or fewer stock than you did ten years ago?

More _____ Fewer _____ Explain why _____

_____

108. Have you changed the way you manage stock compared with ten years ago? Y _____ N _____ Explain: _____

_____

109. Do you have any improved stock? _____ If yes,

110. Specify breed name: _____

111. If respondent is married, ask: "When you were married, what were the terms of your bride-price, that is, amount in shillings, number of cattle, etc.?"

_____

_____

112. What year was that? _____

(Interviewer: Repeat for second and third wives)

_____

_____

113. Approximately, what is the current bride-price? _____

_____

Ujamaa Vijijini

During the past 20 years the government had the program of *ujamaa viji-jini*, whose purpose was to help develop Tanzania's people and productive resources and improve people's lives.

114. In your opinion, what were the effects of the *ujamaa* program?

_____

_____

115. Did good things come from the *ujamaa* program?

Y _____   N _____ . If yes, what were they? _____

_____

116. Did bad things come from the ujamaa program?

Y _____   N _____ . If yes, what were they? _____

_____

Environmental Assessment and Coping Strategies

Farmers have different problems to worry about in different parts of Tanzania. There are dangers of too much rain, too little rain, floods, rains coming at the wrong time, army worms, locusts, *dudu*s, bushpigs, and other animals eating the crop, birds, thieves, changing prices for crops farmers sell, and so forth.

117. For yourself, what are the main risks you face in making a living here?

_____

_____

Of the risks which you have mentioned, which is the

118. Most important? _____

119. Second most important? _____

120. Third most important? _____

Drought Experience

121. What was the worst year for crops that you have experienced here since 1973? _____

122. What caused the failure of the crops? _____

123. If farmer mentions year(s) ask: "If you remember bad years from long ago (such as 1971 and 1972), how does the year(s) you mentioned compare with those years?" _____

_____

124. When the rains have been late, how have you or your family suffered?
_____

125. What did you do at those times? _____
_____

126. Is it possible to influence the amount of rain that falls?
_____

127. If yes, how? _____
128. Can you do anything to help your crops get the moisture they need
to mature? _____

Coping Strategies

I am going to read a list of things a person might do if the rains are late
or not good. Please tell me if you have ever done any of these things. If
you wish, you may give reasons for doing or ***not*** doing any of the things
mentioned.

129.1   Y_____ N_____ Ask help of a kinsman.
129.2   Y_____ N_____ Go to live with a kinsman.
129.3   Y_____ N_____ Send your children to live with a kinsman.
129.4   Y_____ N_____ Move your cattle far away.
129.5   Y_____ N_____ Sell some of your cattle.
129.6   Y_____ N_____ Cultivate *mabonde*.
129.7   Y_____ N_____ Plant a drought-resistant crop.
129.8   Y_____ N_____ Plant more seeds per hole.
129.9   Y_____ N_____ Plant more acreage.
129.10 Y_____ N_____ Thin out the crop standing in the fields.
129.11 Y_____ N_____ Weed more.
129.12 Y_____ N_____ Irrigate.
129.13 Y_____ N_____ Cultivate ridges.
129.14 Y_____ N_____ Stop tending the fields.
129.15 Y_____ N_____ Pray.
129.16 Y_____ N_____ Pay a rainmaker.
129.17 Y_____ N_____ Buy food.
129.18 Y_____ N_____ Hunt or fish.
129.19 Y_____ N_____ Collect food in the bush.
129.20 Y_____ N_____ Try to get a loan.
129.21 Y_____ N_____ Ask for help from District Council.
129.22 Y_____ N_____ Ask for help from cooperative society.

129.23 Y_____ N_____ Seek wage work.

129.24 Y_____ N_____ Do nothing at all.

130. Are there things you have done when the rains are late or have failed that I have not asked about? Y _____ N _____

131. If yes, what are they? _____

_____

_____

Of all the things you have mentioned doing, which do you do

132.1 1st? _____ 132.2 2nd? _____ 132.3 3rd? _____

Of all the things you mentioned doing,

133. which is easiest to do? _____

134. which is hardest to do? _____

135. Do you have any problems of too much rain? _____

136. If yes, explain what the problems are and when they usually occur.

_____

137. When the rains are late, how long can you wait before you become worried about the delay in the beginning of the rains?

_____

Other Income-Producing Occupations

The government is trying to encourage people to find other ways to make a living in addition to farming.

138. Do you or any members of your household have any occupation besides farming and keeping livestock (the things we have talked about so far) that brings you income?

_____

_____

_____

For example, are you or your wife able to improve your livelihood by doing any of the following:

139. Y_____ N_____ carpentry

140. Y_____ N_____ making charcoal

141. Y_____ N_____ making medicine

142. Y_____ N_____ making pots or baskets

143. Y_____ N_____ brewing beer for neighbors

144. Y_____ N_____ cutting and selling wood (kuni)

145. Y_____ N_____ running a *duka*

146. Y_____ N_____ running a tearoom

147. Do you or any of your family ever participate in work parties for other farmers? Y _____   N _____

    If yes, what tasks do you do? _____

_____

The government is encouraging people to do things in a modern way and to use new things in their farming and their lives. How many of the following things have you been able or lucky enough to get?

148.1 Y_____ N_____ plow
148.2 Y_____ N_____ shared use (or hiring) of a tractor
148.3 Y_____ N_____ hand maize mill
148.4 Y_____ N_____ cart
148.5 Y_____ N_____ wheelbarrow
148.6 Y_____ N_____ tilley lamp
148.7 Y_____ N_____ radio
148.8 Y_____ N_____ watch
148.9 Y_____ N_____ *pikipiki*
148.10 Y_____ N_____ bicycle [omitted on Kiswahili version]

149. In your opinion, what are the most important things that government could do to help you improve your agriculture? _____

150. In your opinion, what are the most important things that you can do to improve your agriculture? _____

151. What have you tried to do to improve your agriculture?

_____

_____

Now a couple questions about your feelings about things.

152. What is the most important thing in life? _____

_____

153. Whom can a person trust? _____

_____

NOTE TO INTERVIEWER.   This next question tries to find out farmers' attitudes toward nature. Be serious when you ask the farmer to reply to the question after you have read the story. Even though it is a story, it is not a foolish one, and the farmer's answer will be of great interest. Be sure that you read the story aloud without suggesting by the way you read it or by the tone of your voice which man *you* would agree with!

    Now I want to read you a short story and get your opinion.

    Once upon a time there was a great drought season and it went on for nearly a year. It got so bad that people had almost no food and they had to kill their livestock for food, and some older people even died because of the famine. The people gathered in a large *baraza* to discuss the famine and the rains. Many people spoke, but the people listened especially to three men.

The first man said that there is little they could do about this drought for it was the will of Allah (God). He said the winds, the rain, the sun—all nature is powerful and man is very weak in comparison. They must suffer whatever happens and hope that God will be favorable next year.

The second man said that what was wrong was that the people didn't have a good kind of agriculture. He said that men should try to understand the weather and crops; they should try to understand nature and live with nature as a partner.

The third man said that people in other parts of the world didn't have the problems they were having here because they had solved them. These people in other parts of the world believed they were able to control the land and water and make it possible to grow crops even if the rains were poor. They felt they were superior to nature and could control nature.

If the *baraza* had been held here during such a famine and you had been at the *baraza*, which of the three men would you have agreed with?

155. First man _____ Second man _____ Third man _____

156. Is there anything you want me to know about your farming that
      I have not asked about? If yes, place comment here: _____

_____

End of Interview

157. Respondent's reaction:
      Enthusiastic _____   Withdrawn _____
      Cooperative _____   Hostile _____
      Indifferent _____   Other (specify) _____

Interviewer's comments:

Does the farmer have:
158.1 Y____N____ *bati* roof   158.2 Y____N____cement floor
158.3 Y____N____brick house   158.4 Y____N____roof water catchment
158.5 Y_____N_____ iron casement windows

Interview situation:
159. _____ at home, _____ in the fields, _____farmer alone,
      _____ in group (family), _____ in group (visitors)

160. Check if present: _____ *balozi,* _____ agricultural field
     officer, _____ other official person (specify)
161. Duration of interview: _____ Time finished
                              _____ Time commenced
                              _____ Duration

Interviewer: Go over the whole interview while still with farmer to check
for completeness.

Interviewer: Any spillover from question 34 can go here.

34. (continued)				No. years education completed*	Presently** living here	Does this person work on farm?
Name	Relationship	Sex	Age			
				Y	Y N	Y N
				Y	Y N	Y N
				Y	Y N	Y N
				Y	Y N	Y N
				Y	Y N	Y N

(Use back of questionnaire if necessary.)

* If still attending school, circle the Y on table.

** If not present, query 35.

35. Does he (she) return to help with the farm work? _____
    If yes, specify season: _____

Analysis of the Crop Calendars: Timing of Planting and Season Length for Maize

Kwadundwa, Kwediamba, and Minazini are omitted.

Mgera

MASIKA RAINS. The planting date truly begins on 27 January. The people say they plant on 18 January, which would certainly work out well. The actual planting dates are not greatly dispersed. All of them fall within about four weeks of the first possible planting date. The potential yields for crops planted on 27 January equaled 84.9 percent. Those planted after 27 January averaged 75.3 percent of the potential yield. We are dealing here with reliable yields in any case.

VULI RAINS. They grow maize that takes about 130 days to mature and say they plant on 15 December. This is only one month before the maize for the *masika* rains is planted, so in essence both crops use the same set of rains. Performance is quite good, at 55.8 percent of potential yield, but not as good as that achieved during the *masika* rains. The first possible planting date for the simulation is 13 December. Four of 15 seasons began on 13 December and planting occurred as late as 11 February, but seasons when planting was late gave better results than those planted in mid-December. An early planting date does not guarantee a higher yield.

Planting schedules in the *vuli* period are responsive to work schedules rather than crop calendars.

Kwamsisi

MASIKA RAINS.   The people plant maize on 12 March and the crop matures in 123 days. In half the seasons (of the simulation), the crop could commence that early. Only one season (1977) was a total failure, and the overall level of yield was 74 percent of potential. The planting date seems well chosen.

VULI RAINS.   According to the farmers, they plant maize on 27 September and it takes 107 days to mature. Most of the farmers grew maize during the *vuli* rains. One-third of the time the crop failed altogether in the simulations, and overall only 35.7 percent of the potential yield was achieved. The earliest planting date was 27 September. This is a good choice of date. The farmers might do a little better with a 90-day maize that matured by 10 January rather than the 105-day maize that matures by 25 January.

Mzundu

MASIKA RAINS.   The people say they plant on 25 January. We got 9 instances when the planting date chosen by the computer simulation was 27 January. But in many instances (10 of 32) planting took place in March or early April and one season had no suitable planting date at all (1965). In those seasons when the planting was done on 27 January, the farmers realized 65.6 percent of potential yield. In seasons when the planting took place after 1 March, they realized 73.8 percent of potential yield. The seasons with planting dates between 27 January and 1 March gave 65.5 percent of potential yield. It appears that any time of planting after 25 January works out about the same so long as it catches the April maximum of rainfall.

VULI RAINS.   The farmers say they plant on 10 October and maize takes 141 days to mature. We simulated with 150-day maize. The results were exceedingly poor. In 3 seasons, no suitable planting date occurred, and all but 9 (of 30) seasons had no yield at all. Only 9 percent of potential yield was achieved. The starting dates were all over the place, ranging from 13 October to 21 December. The moisture requirements of a 150-day maize crop are much greater than the rain that comes during the *vuli* season, and the people would be well advised to choose another crop, such as sorghum, or at least a maize variety that matures more quickly. The planting date is well chosen; but the crop itself is not.

Kwamgwe

MASIKA RAINS.   The first possible planting day is 27 February, and there are 7 such years out of 31. The people of Kwamgwe say their maize takes 106 days to mature. They could probably do as well or better by going to a 120- or 150-day crop.

VULI RAINS.   Farmers in Kwamgwe say maize matures in 115 days and they plant on 27 September. We simulated 105-day maize. The planting date appears to be optimally chosen, but even with 105-day maize, in half the seasons the crop failed totally (13 of 28 seasons). A different crop or 90-day maize would likely be a better choice.

Mandera

MASIKA RAINS.   The people claim to plant on 6 March and that their maize takes 134 days to mature. We used 120-day maize in the simulation and started the search for a planting day on 1 March. The earliest day for planting was 13 March. Five of 19 seasons had a 13 March planting date. Only one outright crop failure resulted (1962). The planting date seems well timed and the people achieved 74.5 percent of potential yield.

VULI RAINS.   The farmers said that they planted around 28 October and that maize took 130 days to mature. The simulation gave the earliest planting date as 27 October, and in four seasons that was the planting date. In 1963 the planting date was not until 11 January. Three of the simulated seasons were outright failures, and overall only 7 percent of potential yield was realized. The planting date was not well chosen, and the people should choose a faster-growing, more drought-tolerant crop.

Mlembule

MASIKA RAINS.   The people plant maize on 10 March and it takes 127 days to mature. Although there are occasional total failures (3 in 32 seasons and six years with yields of 35 percent of potential or less), overall the place does pretty well, achieving 56.6 percent of potential yield. The planting date seems well chosen.

VULI RAINS.   Eight farmers said that the *vuli* rains are too unreliable to use. No simulation analysis was done for the *vuli* rains for Mlembule.

Mkomazi

MASIKA RAINS.   It is impossible to grow rain-fed crops in Mkomazi. All crops are irrigated. Maize is generally planted on 15 May and matures in 146 days according to local farmers. Since the crop is irrigated, the timing of planting is related to the flow of water in the Mkomazi River, whose catchment comprises the drainage from the western Usambara Mountains and the eastern Pare Mountains.

VULI RAINS.   Two farmers cited a planting date of 20 October and maturity in 120 days. A simulation based on rainfall gives total failure, but such a late crop may be possible if irrigated. In the eight-month period June through February, only December is not a drought month. Any farming during the *vuli* period must be irrigated.

Vugiri

MASIKA RAINS.   Vugiri has ample rainfall and no problems of moisture shortage. According to the local people, maize is planted 21 February and takes 119 days to mature. This planting date lies on the steep upward curve of rainfall. There is always enough moisture for the crop.

VULI RAINS.   The farmers say they plant maize on 6 September and it matures in 129 days. The date seems well chosen, and in 26 of 58 instances (in the simulation) the crop could be planted on the earliest possible day, 13 September. Over 50 percent of potential yield was achieved with

120-day maize, and 14 seasons were total failures. The *vuli* rains are clearly more risky than the *masika* rains, even in this highland location.

Magoma

MASIKA RAINS.   The people say they grow maize that takes 127 days to mature and they plant on 13 March. They have spotty results. Five of 17 seasons were total failures. Overall they got only 36.1 percent of potential yield. Nonetheless, their planting date was well chosen.

VULI RAINS.   The *vuli* rains are essentially useless for maize cultivation. Nonetheless, eight people gave information on planting dates and time to maturity. They generally planted on 6 October and grew maize that took 112 days. This seems a late date to start and simulation results based on 6 October gave crop failures 80 percent of the time.

Kisiwani

MASIKA RAINS.   Given Kisiwani's greater rainfall and lower evapotranspiration rates, there are even fewer problems here than in Kiwanda (see below).

VULI RAINS.   The farmers in Kisiwani say maize takes 100 days and is planted on 20 September. In planting a crop two weeks earlier than they do in Kiwanda, the farmers of Kisiwani get a distinctly better result: 48.5 percent of potential yield versus 35.6 percent of potential yield, and three total crop failures versus eight. (The simulation used a 120-day maize in both places.) The fact that maize grown in Kisiwani is said to mature some two weeks sooner probably also helps improve crop performance.

Kiwanda

MASIKA RAINS.   Twelve seasons in Kiwanda started on the first possible planting day, 13 March, and most seasons started within four weeks of that date (all but two seasons). Crops planted on 13 March achieved 88.3 percent of potential yields. This is not a place with very serious water-deficit problems, and the date chosen for planting during the *masika* rains is good.

VULI RAINS. The people in Kiwanda say maize takes 124 days and is planted on 6 October. The results are not good. They should plant a couple weeks sooner, the way their neighbors in Kisiwani do.

Daluni

MASIKA RAINS.   The *masika* rains are the lesser of the two seasons. People say they plant maize on 1 April and it matures in 114 days. We ran simulations with the earliest possible planting date of 13 March. In fact, 10 of 21 seasons could start on that day, so perhaps the people of Daluni could do better by planting a couple weeks earlier. Crops planted on 13 March realized 77.8 percent of potential yield. Those planted two weeks later achieved only 53.8 percent of potential yield. In other words, there is risk of a poorer performance in planting at the beginning of April. Figure 5.28 shows that the peak of the *masika* rains is in April, so it is probably better to plant well before the peak.

VULI RAINS.   The *vuli* rains in Daluni are heavier than the *masika* rains, but they are also less reliable. In the period September through December there were 11 instances of a month with no rain at all, whereas in the period March through June there was only 1 instance of a month with no rain at all. Farmers said that they planted on 8 October and maize took 112 days to mature. We simulated using 120-day maize and an earliest planting date of 13 October. A 120-day maize achieved 34.5 percent of potential yield and there were six seasons of outright failure. Farmers in Daluni should choose an earlier planting date, say 15 September, and a faster-growing crop. (Perhaps our simulation did not do the Daluni farmers justice, since our crop grew 8 days longer and was planted at least 5 days after their professed average starting date.)

Mwakijembe

MASIKA RAINS.   We have a very meager record for Mwakijembe, but enough to suggest that this is an exceedingly marginal place to plant rain-fed crops. The people say maize takes 105 days and they plant on 25 April. They could try to plant earlier. Simulations based on this season length and planting date gave abysmal results: six of eight seasons an outright failure and two seasons with yields of 18 and 38 percent of potential. They would be slightly better served by going to 90-day Katumani maize and planting about a month earlier, but even so the results would not be greatly improved. The local people concentrate on livestock and what maize they grow is planted in low areas near the Umba River. The fields receive seepage, so in a way the crop is irrigated.

VULI RAINS.   Five farmers gave data on maize grown during the *vuli* rains. They plant about 28 September and the crop takes 116 days to mature. A 120-day maize planted thus fails completely seven out of nine times. By planting in the *mabonde* areas that receive subsoil drainage, they may do somewhat better than our simulation suggests (table 6.8). In any event, the date chosen for planting is appropriate for the chancy, meager rains of the *vuli* season.

Maranzara

MASIKA RAINS.   The people say their crop matures in 151 days and they plant on 15 March. This date is on the ramp to higher rainfall. The planting dates are well clustered. Forty percent began at the earliest possible planting date and all occurred within four weeks of the first possible date. However, growing a 150-day maize crop poses problems, especially on the sandy, droughty soils of Maranzara. Those seasons with planting on the first possible date (13 March) achieved 57.5 percent of potential yield. Seven late-season plantings (after 1 May) average only 25.4 percent of potential yield, with five seasons of outright failure. The farmers should plant 120-day maize.

VULI RAINS.   Few people seem to grow maize in the *vuli* rains, and our simulation suggests why: 29 of 41 seasons were total failures and only about 10 percent of potential yield was achieved. The date is well chosen but the long-season crop is not.

Moa

MASIKA RAINS.   Only two people claimed to plant maize during the *masika* rains. They said that they planted on 25 March and that maize matured in

96 days. The starting dates in the simulation for Moa are all over the place, covering a nine-week period into early May. The earliest possible planting date was 27 February, but only 5 of 63 seasons started that early. The modal date, in fact, is about 21 March, which is close to the date the two farmers claimed as the day on which they normally planted. With a 90-day maize crop they do very well, getting 85.5 percent of potential yield, but as noted earlier, this is really a fishing village.

VULI RAINS.   Only two farmers said they grew maize during the *vuli* rains, and only one gave planting and maturity dates; thus, we have little information to go on. We did, however, simulate a 105-day maize. We began the search for a planting date on 15 October. The earliest time a farmer could plant was 27 October, and 18 seasons (out of 60) began that day. Overall a farmer would achieve only 9 percent of potential yield, and crops failed two out of three years. Clearly, practically no one grew maize in the *vuli* rains, and the farmers have read the environment correctly. The same can be said of the *mchoo* rains, when crops would be planted in July.

Chronology for Tanga Region

In this appendix I summarize the state of affairs (food, weather, and livelihood) year by year from 1922 to 1975. The summation is created from a close reading of a large number of files from district and provincial offices in Tanga Region and in the Tanzanian National Archives. It has been supplemented by data in Sumra 1975a: 26–28.[1] Not all years are represented. In most instances the locale or district is mentioned. The point will be readily evident that although drought is a common event and causes food shortages, there are numerous other things that affect food supply—in particular, too much rain, badly timed rain, agricultural vermin (bushpigs, baboons, rats, elephants, etc.), locusts and other insects, as well as crop pathogens. An analysis of the chronology is presented in chapter 7.

1922   In Pangani District there were problems with rats and other agricultural vermin. There was an unusual amount of rain. Early fears of rain failure and famine were not borne out. Opportune rains and good crops resulted. There were problems with weevils in maize.

1923   In Pangani District, the rains failed and there was food scarcity.

1924   Insufficient food on sisal plantations prevented the Tanga District commissioner from reporting that the physical condition was "satisfactory or credible. In part ration money is spent on 'wine and women' in lieu of properly cooked food."

1925   The district officer, Pangani, threw 100 bags of purchased maize, bought in anticipation of food scarcity, into the river because there were no purchasers. There was a serious famine throughout Handeni District.

1926   Food shortages were reported in Pongwe and Gombero, Tanga District. Semifamine conditions existed in certain districts from May to October.

1928   In Tanga District, the *vuli* rains were a failure. When rain came, it was excessive but did not last long enough.

1929   The *masika* rains were a great success. There were locust invasions in January and also in February and March in the eastern Usambaras. Serious famine occurred throughout Handeni District.

1932   In the Mwakijembe area (northern Tanga District), there was an almost complete failure of the short (*vuli*) rains. Locust swarms ruined crops in eastern Uzigua (Giblin 1992: 158), leading to famine.

1933   The *masika* rains were extremely unsatisfactory. The rains were delayed in the Digo and Bondei areas of Tanga District. There was drought and locust destruction. There was not a sufficient supply of foodstuffs to feed laborers on the estates. Drought extended through the *mwaka* rains in 1933, leading to a killing famine (Giblin 1992: 158). There were serious food shortages in eastern Handeni from June onward.

1934   There was still drought in the Mwakijembe area. In Handeni, this was the worst period of famine.

1935   Very serious food shortages occurred in eastern Handeni, especially Mazingara Division.

1936   This was a good year; excellent crops were attained.

1937   From an agricultural point of view, the year was a prosperous one.

1941   Food shortages occurred in Kimbe and Negero from June onward; they spread to Chanika and Mswaki in July and lasted until December.

1942   In Pangani District, food shortages were mentioned as well as problems with agricultural vermin. In February there were locust invasions. Serious food shortages happened in Kwekivu; approximately 1,500 people were affected.

1943   Food sales were prohibited in Pangani District "owing to the serious food position." There was an acute shortage of grain.

A telegram from the district commissioner, Pangani, 23 December 1943, stated, in part: "In my opinion the food supplies in these areas should be classed as 'scarce' but not yet 'dangerously scarce.' . . . I record this opinion however, with the utmost possible reserve, as I have now had experience for nearly six consecutive years of the thankless task of dealing with famines."

1944   In Handeni and Korogwe there were very good harvests and no grain shortages, according to some reports. However, Mascarenhas (1973: 1) showed a local acute food shortage in Handeni District. Sumra (1975a: 26) stated that there was a serious famine in Handeni District, especially Chanika and Mswaki. In May it spread to Kwamsisi, Kimba, and Kwekivu. The number of people affected was 13,114. The food situation in Pangani District was quite satisfactory.

1945   In Lushoto Division, the main rains were late (planting occurred only at the end of April). June was unusually dry and resulted in poor yields in nearly all areas. There were fairly severe food shortages in the first few months in northern Lushoto Division. In these areas they cannot grow crops in the long rains because of excessive cold. Food shortages persisted in Mtae, Mlalo and Mlola.

1948   7,000 tons of maize were exported from Handeni.

1949   The crops were an almost total failure in Handeni, and there was famine. In Handeni, "[S]ome people lost as many as 3 successive plantings owing to the unfortunate and unpredictable weather conditions."

1950   "Every part of [Handeni] District is relying on imported meal and the position has deteriorated during the last month as cash reserves for the purchase of food have begun to run out." In February, Kwamsisi was in great physical distress. In March, in the eastern and southeastern parts of Handeni District, "conditions were found to be disappointingly bad." They were also poor in Mgera, as well as in Chanika Zumbeate and on the borders with Korogwe and Pangani Districts.

1951   Both rains were good in Pangani District. There was much trouble from vermin (rats and baboons) as well as excessive rains. In Tanga Province the *vuli* rains were heavy. The long rains (*masika*) "tailed off too early so that the long rain crops that did not shrivel up yielded very much less than had been hoped." The food and

seed position in Chanika and Mgambo (Handeni District) was
bad. "Most of the maize planted at the end of January has died of
being eaten by Army Worm and the people are so busy looking
for food that there is danger of cultivations being neglected."

1952   It was predicted that surplus food stocks from good areas would
barely suffice to feed the people in the bad areas. There was no
surplus to put in grain stores. As of October, food shortages had
begun in Kwamsisi and Mazingara Chiefdoms. The short (*vuli*)
rains generally failed. There were food shortages in Kwamsisi,
Mazingara, and Mkumburu from March. By October, they had
spread to Chanika, Mswaki, and Magamba. Some 40,000 people
were affected.

1953   In January, food shortages spread to five chiefdoms in Handeni
District, including Chanika, Mgambo, and Magamba Chiefdoms;
40,000 people were affected. By February, maize was being
distributed from 62 centers in eight of Handeni's nine chiefdoms
(all except Kwekivu). Two-thirds of the population were affected.
In November, the food shortage in western Handeni worsened.

1954   The May catch crop[2] was harvested. The June rains failed; it
rained 5 mm. The June rains in the Nguu Hills were a complete
failure, and 10,000 people faced a food shortage. In November,
the short (*vuli*) rains were sufficient, and the food shortage was
coming to a close. In Lushoto District the maize harvest was
below average. The long rains failed and crops in Mtae and
Mguashi were a total loss. There were locust outbreaks in
February and March. In Tanga District, there was a lack of rain in
the latter half of the year, threat of locusts, and occurrence of
foot-and-mouth disease, and the rice crop failed. In Handeni
District, the rains generally failed, there was only a small surplus,
and the harvested maize had such high moisture content that it
could not be stored but had to be sold outside the district. There
were food shortages in Mgera and Kwekivu. In April and again in
November there were locusts, but they did little damage. Damage
from grasshoppers in September and October was greater. In
Korogwe District it was so dry that fires on cultivated land (set in
part to kill rodents and other vermin) got out of control and
destroyed 26 buildings in Korogwe Township. Rainfall was below
average; maize yields were below average and inadequate even
for local needs. There were locust infestations, but the damage

was nowhere severe. In Pangani District the rains were poor, the maize and rice crops negligible, and there was food scarcity. There were threats of locusts, and an outbreak of foot-and-mouth disease occurred.

1955 From January to May there were food shortages in Mgera and Mazingara. On the plains of Lushoto District, in February, there were exceptionally heavy rains; from March to June the rains were good. August to October was very dry and the July-planted crops were a failure. In Handeni District there was "one of the largest bumper harvests ever" and well-distributed rain, except in the Nguu Hills. February's rains there were torrential (the heaviest in living memory). "The result was that miles of road and some bridges were washed away and a great deal of castor seed planted in the valley bottoms was destroyed." Kwamsisi had a food shortage. In Korogwe District the *muati* rains failed and there were localized food shortages.

1956 In Lushoto, there was exceptionally bad rainfall and a failure of the short rains. In February, rainfall was exceptionally heavy in the hills; from March through June it was heavier than normal. From July to October it was exceptionally dry.

1957 There were slight food shortages in Mazingara, Mkata, and Manga from February to May.

1958 There were food shortages in Chanika and Manga from February to May.

1959 Kwamsisi Division experienced slight food shortages from January to May.

1961 There was famine throughout Handeni District from January to May. There were isolated food shortages thereafter until December.

1962 This was a dry year: the "driest on record for some estates in Western Usambaras." Drought conditions were experienced from June onward. Torrential rains at the beginning of the year were followed by poor long rains in most areas and a failure of the short rains. Famine relief was given in Handeni and Same. "Pig, baboon, and monkeys have done their usual amount of damage to crops." In Handeni District, there were serious food shortages in Mswaki, Magamba, Mazingara, Mgera, and Kwekivu areas.

1963 Food shortages occurred in western and central Handeni District until March.

1964   There were low rains throughout Tanga Region. The maize crop
acreage was down in Korogwe and Handeni. In Korogwe, after
the *vuli* rains, the rains essentially stopped until April, which had
a detrimental effect on rice. Lushoto's crops were average to
below average, although the *masika* crop in Lushoto District was
above average.

1965   There was scarce rainfall and it was badly distributed in Handeni
District. The rains were below average. Mwakijembe experienced
some food shortages, although there was a good supply of cassava,
which greatly alleviated the situation. Handeni District had a
difficult year. Some farmers "had to plant three times before
obtaining satisfactory germination and then getting very little
rainfall afterwards to grow the crop on." In Lushoto District, the
rains were below average and there were local maize shortages.

1966   Although the *masika* season rainfall was badly distributed, and in
some places in Handeni District farmers had to replant three
times, the crop season was successful, including a good cotton
crop.

1967   Initially there was unfavorable weather in Tanga Region, but
production during the *masika* rains was plentiful, although the
rains were later than usual. The rice crop was damaged by these
rains. The *vuli* rains were good this year. Until May, central and
western Handeni had serious food shortages; they were slight in
eastern Handeni. The number of people affected was 41,850.

1968   In Tanga Region, there was too much rain. In November, stored
foodstuffs were reported to have lots of insects in them.

1969   The food situation was satisfactory except in Sindeni, Kwamsala,
Kwamkono, and Misima areas of Handeni District, where food
shortages were reported. April was dry and the extreme northern
part of Tanga District continued to have insufficient rain.

1970   There were serious food shortages in eastern Handeni, especially
toward the end of the year.

1971   Food shortages spread to Handeni, affecting 54,599 people. There
was little rain. Hunger and food shortages were reported in
Korogwe District. The government was able to cope with minor
food shortages up until October. In November, hot sun and very
strong seasonal winds brought hunger to the whole district. Food
aid was required. Government blundering increased the suffering
of the people. By January 1972, 31,679 people had consumed

1,766 bags of famine relief food. Overall, a total of 43,000 people were helped. Kwamsisi became dry; cows, goats, and sheep died because of a lack of water and graze. There were major problems with the cotton crop. In this period the people of Ngomeni had a hard time cooperating and sticking together. The 1971 *vuli* rains were very poor and food aid was needed in eastern Handeni and Pangani Districts. Tanga District did not require food aid.

1972   In January, food shortages spread throughout Handeni District. The number of people affected was 95,000. Although food shortages eased in western Handeni toward the end of June, they continued in central and eastern Handeni in the second half of the year, affecting 55,000 people. In Pangani, there was little rain in February; crops experienced drought. In October the weather was dry. In Korogwe in May there was heavy rain, which destroyed sweet potatoes and other field crops. In Mombo in January it was dry and windy; there was as a consequence a lot of dust and the crops and graze wilted completely. In Magoma (Korogwe) the sun was very hot, and there were fierce winds causing much dust.

1973   Rains were adequate and crops were generally good.

1974   In April there was too much rain, making it difficult for crops to ripen. It was cold and continued rainy in June. Later on the rains diminished and were poorly distributed. There was a severe drought over much of Korogwe District.

1975   In Pangani District during the *vuli* season it was hot, but rains came and crops like cashews and coconuts ripened.

Notes

Chapter One

1. A drought month is a month with a 75 percent probability of receiving less than 50 mm of precipitation.

2. The climatological data on which my analysis is based are probably unique in their detail and comprehensiveness. The compilation of the data was a long and exhausting task.

3. The archives, of course, reflect an official governmental view of conditions and events. Although interpretations reflect a European bias in the period before 1961, and therefore may be suspect, the facts reported in archival documents can generally be relied upon. A further feature is that files are often fragmentary and incomplete. Some files have been lost or destroyed. For example, in the Tanga Region Archives, files from the various districts of Tanga Region had been brought together. They were bundled with sisal twine and formed a pile 6 feet high in the corner of one room. Many of them, when opened, showed that they had been eaten by termites.

4. Other themes to which this research relates are environmental history and African hunger, although in less direct ways.

5. Confusion can result when different researchers invent acronyms for their favored methods. Among the risk-hazard vulnerability community, PAR means "pressure and release," whereas for Robert Chambers, PAR means "participatory action research" (Blaikie et al. 1994: 21; Turner et al. 2003a: 8074; Chambers 1997: 106).

Chapter Two

1. Asking these questions generated some poignant moments. One person observed, sadly, that although he owned a watch, it no longer worked; and another

lamented that his bicycle was useless, because he could not afford to buy the parts it needed. It is notable that in high, rugged country (Vugiri, Kiwanda, Kisiwani), bicycles were much less common, even though a number of people could probably have afforded to buy one.

Chapter Three

1. I focus on precipitation in this study. Except at high elevations, temperatures are warm throughout the year, with monthly means ranging between 23°C (74°F) in the low-sun period (June–July) and 27°C (81°F) in the high-sun period (January–February). For every thousand meters of elevation, one can reduce average monthly temperatures by about 4.9°C. Relative humidity is high in the rainy season and lower in cooler, dry periods.

2. We thank the following for their help with soil information and with ethnographic data on each village: Bakari Mselem (Mzundu), Mbarak Mndalika (Mgera), Ali Mwarwavu (Kwamgwe), Mbelwa Mahaba (Maranzara/Pongwe), Yohana Juma (Kiwanda), Bakari Mohamed (Kisiwani), Jabir Rashid (Moa), Mohamed Ndungo (Mwakijembe), Rashid Zayumba and Gilbert Mahange (Magoma), Musa Omari (Mandera), Charles Abed Daffa (Vugiri), Ibrahim Rashid (Mlembule), Shaban Ngaga (Mkomazi), Abdallah Singonde (Daluni), and Kitambi Hakundwa (Kwamsisi).

3. Oxisols, although easily worked, are low in cation exchange capacity (CEC); that is, they have a low ability to store minerals (such as calcium, magnesium, potassium, and phosphorus) and make them available to growing plants. They also have a lower ability to hold moisture. Alfisols have a greater clay fraction and, thus, a higher CEC; they usually can hold a greater amount of moisture in the soil and, thus, on two counts (soil nutrients and water availability) are better than Oxisols. Unfortunately, Oxisols and Alfisols are similar in color and texture and difficult to distinguish in the field. Ferralitic Arenosols are deeply weathered, well-drained, and high in iron and aluminum oxides and sometimes have concretionary ironstone bands or clay layers. Orthic and rhodic Ferralsols typically are deep red sandy clays to clays with moderate organic matter and are well drained. Orthic Luvisols and Fluvisols are imperfectly to poorly drained, dark gray or gray brown alluvial clays, sandy clays, and clay loams. Entisols are commonly young soils that do not exhibit a well-developed profile. They vary greatly in origins and suitability for cultivation, ranging from highly fertile alluvial soils to acid sandy soils (e.g., former beach terraces) to rocky Lithosols (eutric Regosols) of low fertility. Vertisols, the famous "black cotton soils," develop on low-lying, seasonally flooded areas. Chromic Vertisols are moderately well to imperfectly drained, usually calcareous, black, dark gray, or brown cracking clays. In the dry season, these clay soils form deep polygonal cracks. Although generally fertile, Vertisols are difficult to work, especially when

wet, and they lack phosphorus and nitrogen (Coleman 1996; National Research Council 1982).

4. In writings about Handeni District, one encounters references to both the Nguu Mountains and the Nguru Mountains. They are one and the same, although the main Nguru Mountains are found in northern Morogoro and Kilosa Districts.

5. Following the Latin name, the Kiswahili vernacular equivalent is given in parentheses.

6. In preparing this study I also used work on soil moisture by Dewan (1983), Hathout and Sumra (1974); on soils and physiography by de Pauw (1984); on vegetation by Iversen (1991); on agrometeorological modeling by Frere and Popov (1979), Gommes (1985), Gommes and Houssian (1982), Mhita (1984), Mhita and Nassib (1987); and a suite of reports by the German Agency for Technical Cooperation (1975, 1976a, b, and c).

7. Some of the sources used in matching vernacular names with Latin names of plants are Dale and Greenway 1961; Johnson 1939; Ruffo 1989; Ruffo, Mwasha, and Mmari 1989; and handwritten notes in the Pangani District Book, *Important Trees or Tree Products Found in Pangani District,* undated, reel 3, Tanzanian National Archives.

8. Difficulties in reading handwritten entries in the questionnaires frequently led to uncertainty as to Kiswahili spelling and thus identification of items.

Chapter Four

1. At least four kinds of slave status were recognized: *shamba,* one who worked in fields; *ijara,* one who paid the owner tribute; *kibarua,* one who worked for a daily wage; and *fundi,* a skilled craftsman. At each step along this continuum, the slave had greater autonomy and rights relating to forming a family, holding property, and allocating labor. At one extreme was the newly enslaved, called *mshenzi,* "barbarian," "uncultured," and "despised"; at the other were descendants who were *mzalia,* literally "born here" (Glassman 1991: 288–294).

2. This section borrows from Porter 1976.

3. The expansion of the fly belt was accompanied by large-scale outbreaks of sleeping sickness in the western and southern parts of Tanganyika in the period 1900 to the 1930s, and hundreds of thousands of people were forced to vacate areas into which the fly was advancing.

4. The following sources were used for analysis and in making figs. 4.1 and 4.2: Schnee 1920: 3.292–294; Reiches-Kolonialamt 1914; Potts 1937; Survey Division 1956.

5. The following sources were used in compiling fig. 4.4: Bureau of Statistics 1992b: 68; Sayer 1930: 219; Berry 1971: 172; British Foreign Office 1920b; Mpigachapa wa Serikali 1990: 161; Economics and Statistics Division 1961: 63;

Department of Agriculture 1945; Barclays Bank D.C.O. 1965; Central Statistical Bureau 1968: 83; Ministry of Agriculture and Cooperative Development 1961: 1.32; East Africa High Commission 1961: 81; United Republic of Tanzania 1969a: 47, 1972: 71–72; International Bank for Reconstruction and Development 1960.

6. There is a large literature on *ujamaa vijijini*. I found the following the most useful: Freyhold 1979; Collier, Radwan, and Wangwe 1986; Coulson 1979; Sumra 1975a, b; Raikes 1972. Note also is made of Merten 1989; Rubin 1985; Scheinman 1986; Seppälä 1998; von Mitzlaff 1988.

7. Block farms are large, centrally managed corporate or cooperative farms. They feature large fields, monocropping, and mechanized and technical inputs (tractors, combines, fertilizers, and improved seeds).

8. The following sources were used in making figs. 4.9–4.11: Bureau of Statistics—Takwimu 1992; United Republic of Tanzania 1992: 286–383; Bureau of Statistics 1969.

Chapter Five

1. My description of the 18 villages lacks one important feature: a sense of the dynamics of village governance. We were not in these villages long enough to learn this. Michaela von Freyhold's *Ujamaa Villages in Tanzania* (1979) contains four case studies that give a vivid sense of social structure, leadership, and governance.

2. After independence (1961), TANU instituted a political structure designed to reach and organize the people. The lowest political rank was the ten-cell leader. Ten families reported to a ten-cell leader, who reported to the next highest level in the TANU hierarchy, the village executive officer.

Chapter Six

1. The Culture and Ecology team worked among eight communities in four East African cultures, one farming and one pastoral community in each culture. I wish to thank the following for their help, intellectual stimulation, and generosity: Walter Goldschmidt, project director, and his wife, Gale, in Sebei District, Uganda; Francis Conant, among the Pokot in Kapenguria District, Kenya; Chad and Beje Oliver, among the Akamba in Machakos District, Kenya; and Bud and Patty Winans, among the Wahehe in Iringa and Mufindi Districts, Tanzania. I also thank Robert Edgerton, who administered eight attitudinal and psychological tests to samples of 30 adult men and women in the communities we studied (Edgerton 1971).

2. Much of the research in the soil sciences has concerned fertilizer, soil chemistry, and crop nutrition. Less work has been done on crop-water requirements. Soil moisture is the main focus of this study.

3. That is, dates on which to start the search for a good day to plant.

4. More research has been done on maize than on any other grain crop in East Africa. For example, Acland's excellent book *East African Crops* (1971) cites 42 studies in the section on maize; the brief sections on bulrush millet and finger millet have none. A further problem relates to the calibration of the energy-water balance model, that is, in linking its estimates to trials where the actual yield is known. We made a substantial effort, devoting days of research time, in an attempt to find records at Mlingano Agricultural Research Institute. Our effort was inconclusive and only a paltry set of yields and phenologies was gathered. In the end, I did not have sufficient data to calibrate the energy-water-crop model, and we have only the calibration from John Corbett's dissertation (1990).

5. It should be noted that our simulations throughout this chapter give only *estimated* yields; they are not observed yields.

6. In discussing performance of simulations, I am counting, not villages, but individual rainfall stations, of which there are 16 (two of the stations each serve two villages).

7. The record for Mwakijembe is for only nine crop seasons, but this is sufficiently long to suggest that it is a risky place to grow maize.

8. These values are based on the top 12 seasons, where yields were 46 percent or greater. The overall average for crop season transpiration is 215 mm, but this counts a host of failed seasons, when plants died and had no opportunity to transpire.

9. The figures used here may be a bare minimum for subsistence. Sumra noted that maize is sometimes used to buy other foods (with lower caloric value) and nonfood items, like salt, kerosene, and clothes (Sumra 1975a: 123). He used 300 kg/person, which would give a household target of about 2,000 kg.

10. There is a long tradition in the western Usambara Mountains of irrigating "market garden" crops (beans, peas, broccoli, etc.), which are exported to Dar es Salaam and other urban markets (Heijnen and Kreysler 1971).

Chapter Seven

1. In the first litany I quote from the following archival documents: *Annual Report,* District Commissioner, Tanga, 1929, p. 20; *District Report,* Tanga, 1934, pp. 1, 8; 1/20/31, 24 February 1950, R. Thorne, DC, Handeni, to PC, Tanga, *Report on the Food Shortage, Handeni*, p. 90; 1/20/278, 401, 14 June 1950, R. Thorne, DC, Tanga, to the Regional Asst. Dr. of Ag., Tanga and Northern Province, Arusha, *Recurrent Food Shortages in Handeni*, p. 128; 1/20/278, 401, 7 September 1950, R. Thorne, DC, Handeni, to PC, Tanga, *Famine Expenditures*, p. 148; *Annual Report,* Pangani District, 1923, p. 7; file 4/441/I, ref. 122/1, Grain Storage, Department of Agriculture, Moshi, 29 April 1940, p. 1.

In the second litany I quote from the following archival documents: 1/20/31, 24 February 1950, R. Thorne, DC, Handeni, to PC, Tanga, *Report on the Food Shortage, Handeni*, pp. 88, 91, 92; Tanga Province, *Provincial Annual Report*, 1955, p. 29; Tanga Province, *Annual Report*, 1951, p. 11; DC, Handeni, to PC, Tanga, 4 May 1953, 1/8/532, 4/411/I, p. 233; District Commissioner, Tanga, *Annual Reports*, 1923, 4/411/I, p. 11; 4/411/I, *Notes on Handeni Food Shortage*, 10 February 1953, p. 177; DO, Pangani, to DC, Tanga, 1929, p. 12.

2. 1/20/31, 24 February 1950, R. Thorne, DC, Handeni, to PC, Tanga, *Report on the Food Shortage, Handeni*, p. 92.

3. 4/411/I, 31 January 1953, p. 160.

4. Ref. 166, March 1953, D. Brokensha, DC, Handeni, *Grain Movements*.

5. 110/VI/91, 4/411/I, p. 168, 18 February 1953, Harry Collings, Director of Grain Storage Department, to District Traffic Superintendent, EAR&H, Tanga.

6. 4/411/I, p. 177, 10 February 1953, *Note on Handeni Food Shortage*.

7. DGS 110/VI/13, 4/411/I, p. 171, 28 February 1953.

8. 1/8/435, 4/411/I, p. 175, 4 March 1953, DC, Handeni, to PC, Tanga, *Food Shortage Handeni District*.

9. DGS 110/VI/198, 4/411/I, p. 186, 5 March 1953, G. H. Rulf, Dir. of Grain Storage, to Traffic Superintendent, EAR&H, Tanga.

10. Ref. 411/I, 7 March 1953; ref. 1/8A/21, 13 March 1953, DC, Handeni, to DC, Lushoto; 411/1/221, 4/411/I, p. 221, 10 April 1953, PC, Tanga, to Secretariat, DSM.

11. 411/1/221, 4/411/I, p. 221, 10 April 1953, PC, Tanga, to Secretariat, DSM.

12. *Annual Report*, Handeni District, 1954, p. 36.

13. Ibid., pp. 36, 42.

14. File 4/411/I, ref. 122/1, Grain Storage, Department of Agri., Moshi, 29 April 1940.

15. Ref. 1/14/52, 25 April 1950, R. Thorne, DC, Handeni, to Director, Grain Stores Department, Moshi.

16. Ref. 1/14/67, 14 September 1950, R. Thorne, DC, to PC, Tanga Province.

17. Ref. 1/1488, 1 February 1951, R. Thorne, DC, Handeni, to DC, Lushoto.

18. Ref. 1/14/121, 2 May 1951, Thorne, DC, Handeni, to Director of Grain Storage, Moshi; ref. 10/13/1481, 7 May 1951, Regional Assistant, Director of Agriculture, to Entomologist; ref. 14/1/149, 8 January 1952, T. R. Sadleir, DC, Handeni, to PC, Tanga.

19. Ref. 4/411/I, pp. 26–29.

20. Ibid., p. 27, italics in original.

21. Ibid.

22. Ibid., p. 29.

23. Sumra (1975a: 77) listed 10 wild fruits, 6 wild plants whose leaves can be eaten, and 4 plants whose roots or tubers can be or are consumed in times of severe hunger. We have identified the following: fruits—*mavumo* (*Borassus flabellifer*, a palm) and *malanga* (*Grewia plagiophylla*); leaves—*mchicha* (*Amaranthus* spp.),

moronge (*Moringa pterigosperma,* horseradish tree), and *mkunungu* (*Terminalia catappa,* Indian almond tree); and roots—*ndiga* (*Dioscorea dumetorum*) and *mdugu* (*mdudu; Thylachium africanum*).

24. There is a large anthropological literature dealing with attitudes of the individual and the group toward nature, environmental risk, and natural hazards. Mary Douglas has contributed greatly to this theme (Douglas 1963, 1985, 1992; Douglas and Wildavsky 1982). Although she and her colleagues developed a complex codification of terminology, I prefer the classification developed for the story told the farmers, because it well represents three distinct views of individual and societal relationships with nature. As Douglas herself observed: "[T]he risk taking facet and the risk-averse facet of the personality each emerges in the course of a public debate on freedom and control. It is not psychology but anthropology that shows how a community forces sharp and clear ideas about the self on its members" (Douglas and Calvez 1990: 445).

Her terminology flows from a two-by-two table. On the *x*-axis is an individualized–collectivized continuum; on the *y*-axis is a prescribed–prescribing continuum (i.e., the degree to which there are externally imposed restrictions placed on choice). From this are derived four views of nature: *nature capricious* (the fatalistic individual; upper-left box), *nature perverse/tolerant* (the hierarchist, who relies on organization; upper-right box), *nature benign* (the individualist, an economic rationalist; lower-left box), and *nature ephemeral* (the egalitarian, who seeks by critical analysis to live harmoniously with nature; lower-right box) (Schwarz and Thompson 1990: 9).

25. In this discussion I use the term "Tanga District" because that was the name of the territory until 1975, when it was reconfigured and divided into two districts—Muheza and Tanga (peri-urban).

Chapter Eight

1. Dr. Samuel E. Mugogo, director of MARI, kindly provided us with office space, a place to plug in our laptop computer, and access to MARI publications and staff during our time in Tanga Region. We greatly appreciated his help and interest.

2. Our research design had one flaw of which I was not sufficiently aware at the time the study was planned. Geographers generally study places, and these places are usually peopled. Anthropologists study people, who happen to live in places. I chose to revisit a randomly sampled population in the same 17 places 21 years later. This is the same population in the same places, but the two-thirds who had survived were all 21 years older (and so was I). Some of them had largely retired from farming, some were senile, and a few were deaf or blind. Thus, when a head of household answered that he (or she) had stopped growing cowpeas, was it because

he now farmed less than he had formerly, or was this a change in his practice? There are advantages in seeing how practices and attitudes have changed, but it has been tricky knowing when to attribute them to general societal change and when to a person's entry into a different age-group. At times, I wished we had returned to the same villages but drawn a new sample. Yet, clearly, we benefited by visiting the same people again, adding on to their life histories, and seeing how their views and answers had changed. Further, the strong similarity between the age-sex pyramids we obtained in 1972 (fig. 4.9) and 1993 (fig. 4.10) is reassuring. It suggests that we were interviewing a representative, comparable sample of the population of the villages, even though the choice of those to be interviewed (about a third of them) was determined by blood, not by chance.

3. I am reminded of the suspicion farmers on Mount Kilimanjaro exhibited when colonial officials showed up offering to carry out cadastral surveys, leading to registration of farmers' land. In an example of clear thinking, the farmers replied, "If we accept this title to our land, we admit that it is not ours but that it belongs ultimately to those from whom we receive the title" (Allan 1965: 407).

4. Confusion can result when different researchers invent acronyms for their favored methods. For Robert Chambers, PAR means "participatory action research," whereas among the risk-hazard vulnerability community, PAR means "pressure and release" (Blaikie et al. 1994: 21; Turner et al. 2003a: 8074).

Appendix D

1. The chronology was compiled from an analysis of the following documents (arranged by date) in the Tanzanian National Archives (Nyaraka wa Taifa) and the Tanga Region Archives. Sumra's work (1975a) was most helpful in preparing the chronology. He had access to Agricultural Field Officer Monthly and Annual Reports (1950–1972) and Annual Reports of the Director of Agriculture (1923–1935). At the time of our research, many files from Handeni District (and other Tanga Region districts) existed in the Tanga Region Archives in the form of numbered bundles, tied with sisal twine, and these were not available to us.

Annual Report, Pangani, 1922, p. I.

Annual Report, Pangani, 1923, pp. v, vii.

Annual Reports, District Commissioner, Tanga, 1923, p. 11.

Annual Report on the Tanga District for 1924, pp. 1, 11.

Annual Reports, District Commissioner, Tanga, 1925, p. 6.

Annual Reports, District Commissioner, Tanga, 1926, pp. 13–15.

Ref. 252/30, *Annual Report on the Tanga District for the Year 1928*, 20 February 1929.

Annual Reports, District Commissioner, Tanga, 1929, pp. 1, 12, 17, 21.

Acc. 31, file 40, DO, Pangani, to PC, Tanga, n.d. (c. 1929).

District Officer's Report, Tanga, 1931, p. 10.

District Officer's Report, Tanga, 1932, p. 7, Agriculture, 8 March 1932.

District Officer's Report, Tanga, 1933, p. 19.

Acc. 31, file 40, PC (Longland), Tanga, to DO, Pangani, 1 July 1933.

District Report, Tanga, 1934, pp. 1, 8, 10–12.

Report on Native Affairs, Tanga, 1936, p. 1.

Report on Native Affairs, Tanga, 1937, p. 1.

Acces. 1071/1, Soil Erosion, Ag. Officer, Northeastern Circle, Moshi, to PC, Tanga, April 1940.

4/411/I, ref. 122/1, Department of Agriculture, Moshi, 29 April 1940.

Acc. 31, file 40, *Pangani Handing Over Notes*, February 1942, p. 2.

Acc. 31, file 40, *Pangani Handing Over Notes*, April 1942, p. 5.

1610/2 4/411/I, R. May, DC, Tanga, to PC, Tanga, 28 July 1943, p. 6.

4/411/I, ref. 65/4/209, DC, Pangani, to PC, Tanga, n.d. (c. 1943).

Acc. 31, file 40, telegram, DC, Pangani, 23 December 1943.

Agricultural Office, Lushoto, 28 December 1945, pp. 3, 13.

1/20/31, R. Thorne, DC, to Prov. Com., Tanga, 24 February 1950, pp. 88, 91.

1/20/245, *Monthly Report on the Food Shortage*, n.d. (c. March 1950), p. 113.

410 10/2/1137, A. H. Savile, Prov. Ag. Officer, to Dir. of Ag., DSM, 1 April 1950.

4/411/I, ref. 1/14/52, R. Thorne, Handeni, to Director, Grain Stores Department, Moshi, 25 April 1950.

File 21/1, Relief, Korogwe, DO, Korogwe, to Manager, Ngomhezi Estate, 27 December 1951.

File 962 (vol. 19), *Annual Report*, 1951, Pangani District, pp. 5, 7.

Tanga Province Annual Report, 1951, pp. 5, 11.

Ref. no. 411/140, [no title], July 1952.

1/8/196 4/411/I, DC, Handeni, to PC, Tanga, *Food Shortage*, Handeni District, 7 January 1953, pp. 156, 160, 168, 171, 175, 177, 186.

Ref. 44/175, Provincial Produce Officer, Tanga, to PC, Tanga, 21 March 1953.

28584/52 4/411/I, Member for Social Services, DSM, to PC, 15 April 1953, p. 226.

1/8/532 4/411/I, DC, Handeni, to PC, Tanga, *Food Shortage Handeni District*, 4 May 1953, p. 233.

232/15 4/411/I, Senior Medical Officer to PC, Tanga, 5 May 1953, p. 235.

Ref. 1/14/177, DC, Handeni, to PC, Tanga, 6 August 1952.

Ref. 1/14/214, DC, Handeni, to PC, Tanga, 17 October 1953.

Ref. 1/14/162, DC, Handeni, to Director of Grain Storage, 26 November 1953.

Ref. 1/8B/2, DC, Handeni, to PC, Tanga, 31 December 1953.

Ref. 1/8/617, *Food Shortage Handeni District*, June 1953, 4 July 1953.

Ref. 411/1/277, Acting PC, Tanga, to Member for Local Government, 1 September 1953, p. 217.

Ref. 411/1/279, to Member for Local Government from J. C. Claver, 3 September 1953.

Ref. 1/8/48, Political [Officer], Handeni, to Provincer, Tanga, 5 June 1954, p. 346.

Ref. 411/3355, PC, Tanga, to the Honourable, the Chief Secretary, to the Government, the Secretariat, Dar es Salaam, 13 July 1954.

Saving 1/8/40, Political [Officer], Handeni, to Provincer, Tanga, 3 December 1954, p. 308.

File 962 (vol. 19), *Annual Report*, 1954, Pangani District, pp. 7, 37, 43, 45, 73, 80.

File 962 (vol. 19), *Annual Report*, Lushoto, 1954, pp. 91, 106–107.

Annual Report, Pangani, 1954, p. 172.

Annual Report on Agriculture, Lushoto District, 1954, pp. 2, 6.

Annual Report on Agriculture, Lushoto, 1955, p. 3.

File 304/962, *Tanga Province Annual Report 1955*, p. 11.

File 304/962, Handeni District, 1955, pp. 58–59, 65.

Annual Report, Tanga, 1964, p. 20.

Acc. 481, file A3/2, vol. 3, *Monthly Report*, Tanga Region, August 1968, p. 9.

Acc. 481, file A3/2, vol. 3, *Monthly Report*, Tanga Region, October 1968, pp. 11, 13.

Acc. 481, file A3/2, vol. 3, *Monthly Report*, Tanga Region, November 1968, p. 10.

Acc. 481, file A3/2, vol. 3, *Monthly Reports, Agriculture*, Tanga, September 1969, pp. 42–43.

Taarifa ya Ugawaji ya Chakula katika Sehemu Zilizokkukuwa na Upungufu, Korogwe, 15 December 1971.

2. A catch crop is a crop that can be grown quickly, to provide food when it is scarce.

Bibliography

Acland, J. D. 1971. *East African Crops.* London: Longman.

Acock, B. 1989. "Crop modeling in the USA." *Acta Horticulture* 248:365–371.

Ainley, John. 2001. *Pink Stripes and Obedient Servants: An Agriculturalist in Tanganyika.* Driffield, East Yorkshire, UK: Ridings Publishing Co.

Akehurst, B. C., and A. Sreedharan. 1965. "Time of planting—a review of experimental work in Tanganyika, 1956–62." *East African Agricultural and Forestry Journal* 31:189–201.

Allan, W. 1965. *The African Husbandman.* Edinburgh: Oliver and Boyd.

Atlas of Tanganyika. 1956. 3rd ed. Dar es Salaam: Survey Division, Department of Lands and Surveys.

Atlas of Tanzania. 1976. Dar es Salaam: Surveys and Mapping Division, Ministry of Lands, Housing, and Urban Development.

Bahemuka, J. M., and M. Tiffen. 1992. "Akamba institutions and development, 1930–1990." In M. Tiffen, ed., *Environmental Change and Dryland Management in Machakos District, Kenya, 1930–90, Institutional Profile.* Working Paper. London: Overseas Development Institute.

Baier, W. 1979. "Note on the terminology of crop-weather models." *Agricultural Meteorology* 20:137–145.

Baker, E. C. 1934. *Report on Social and Economic Conditions in the Tanga Province.* Dar es Salaam : The Government Printer.

———. 1949. "Notes on the history of the Wasegeju." *Tanganyika Notes and Records* 27:16–39.

Barclays Bank D.C.O. 1965. *Tanzania.* London: Barclays Bank D.C.O.

Bassett, T. J., and K. S. Zimmerer. 2003. "Cultural ecology." In G. L. Gaile and C. J. Willmott, eds., *Geography in America at the Dawn of the 21st Century.* New York: Oxford University Press.

Batterbury, S., and T. Forsyth. 1999. "Fighting back: Human adaptations in marginal environments." *Environment* 41:6–11, 25–30.

Baumann, Oskar. 1891. *Usambara und seine Nachbargebiete.* Berlin: D. Reimer.

———. 1896. "Der unterlauf des Pangani-Flusses." *Petermanns Geographische Mitteilungen* 42:59–62.

Bax, S. Napier. 1943. "Notes on the presence of tsetse fly, between 1857 and 1915, in the Dar es Salaam area." *Tanganyika Notes and Records* 16:33–48.

Bebbington, A. 1994. "Theory and relevance in indigenous agriculture: Knowledge, agency and organization." In D. Booth, ed., *Rethinking Social Development: Theory, Research and Practice.* Harlow, UK: Longmans.

————. 1996. "Movements, modernizations and markets: Indigenous organizations and agrarian strategies in Ecuador." In R. Peet and M. Watts, eds., *Liberation Ecologies: Environment, Development, Social Movements.* London: Routledge.

Belmans, C., J. G. Wesseling, and R. A. Feddes. 1983. "Simulation model of the water balance of a cropped soil: SWATRE." *Journal of Hydrology* 63:271–286.

Berry, L. 1971. *Tanzania in Maps.* London: University of London Press.

Berry, L., and R. W. Kates. 1970. *Planned Irrigated Settlements: A Study of Four Villages in Dodoma and Singida Regions.* BRALUP Research Paper no. 10. Dar es Salaam: University of Dar es Salaam.

Biermann, Werner, and Jumanne Wagao. 1986. "The quest for adjustment and the IMF, 1980–1986." *African Studies Review* 29:89–103.

Blackie, J. R., and L. Bjorking. 1968. "Lysimeter study of the water use of sugar at Arusha Chini." Paper presented at Fourth Specialist Meeting on Applied Meteorology, Nairobi, 26–27 November.

Blaikie, P. 1985. *The Political Economy of Soil Erosion in Developing Countries.* London: Longman.

————. 1998. "A review of political ecology: Issues, epistemology, and analytical narratives." *Zeitschrift für Wirtschaftsgeographie* 3–4:131–147.

Blaikie, P., and H. C. Brookfield. 1987. *Land Degradation and Society.* London: Methuen.

Blaikie, P., T. Cannon, I. Davis, and B. Wisner. 1994. *At Risk: Natural Hazards, People's Vulnerability and Disasters.* London: Routledge.

Boserup, E. 1965. *The Conditions of Agricultural Growth.* Chicago: Aldine.

British Foreign Office. 1920a. *The Partition of Africa.* Peace Handbooks, vol. 18, no. 89. London: H. M. Stationery Office.

————. 1920b. *Tanganyika (German East Africa).* Peace Handbooks, vol. 18, no. 113. London: H. M. Stationery Office.

————. 1920c. *Treatment of Natives in the German Colonies.* Peace Handbooks, vol. 18, no. 114. London: H. M. Stationery Office.

Brokensha, David. 1971. "Handeni revisited." *African Affairs* 70(279):159–168.

Brokensha, David, D. M. Warren, and Oswald Werner, eds. 1977. *Indigenous Knowledge Systems and Development.* Washington, DC: University Press of America.

Brown, J. M. 1963. "Water use." In H. E. King, ed., *Progress Reports from Experiment Stations, Western Cotton Growing Areas.* London: Cotton Research Corp.

Bruntland, G., ed. 1987. *Our Common Future.* World Commission on Environment and Development. New York: Oxford University Press.

Bryceson, Deborah Fahy. 1990. *Food Insecurity and the Social Division of Labor in Tanzania, 1919–85.* New York: St. Martin's Press.

Bureau of Statistics. 1969. *1967 Population Census.* Vol. 1, *Statistics for Enumeration Areas.* Dar es Salaam: Bureau of Statistics.

———. 1992a. *Agricultural Sample Survey of Tanzania Mainland, 1989/90.* Dar es Salaam: Bureau of Statistics, President's Office, Planning Commission.

———. 1992b. *Agriculture Statistics 1989.* Dar es Salaam: Bureau of Statistics, President's Office, Planning Commission.

Bureau of Statistics—Takwimu. 1992. *Tanzania Sensa 1988, Population Census, Regional Profile, Tanga.* Dar es Salaam: Bureau of Statistics.

Burton, I., and R. W. Kates. 1964. "The floodplains and the seashore." *Geographical Review* 54:366–385.

Butzer, K. W. 1989. "Cultural ecology." In G. L. Gaile and C. J. Willmott, eds., *Geography in America.* Columbus, OH: Merrill Publishing Co.

Carney, J. 1993. "Converting the wetlands, engendering the environment: The intersection of gender with agrarian change in The Gambia." *Economic Geography* 69:329–348.

Carr, M. K. V. 1968. "Report on research into the water requirements of tea in East Africa." Paper presented at Fourth Specialist Meeting on Applied Meteorology, Nairobi, 26–27 November.

Cash, D. W., W. C. Clark, F. Alcock, N. M. Dickson, N. Eckley, D. H. Guston, J. Jäger, and R. B. Mitchell. 2003. "Knowledge systems for sustainable development." *Proceedings of the National Academy of Science* 100:8086–8091.

Central Statistical Bureau, United Republic of Tanzania. 1967. *Directory of Industries, 1967.* Dar es Salaam: The Government Printer.

———. 1968. *Statistical Abstract, 1966.* Dar es Salaam: The Government Printer.

Chaiken, Miriam S., and Anne K. Fleuret, eds. 1990. *Social Change and Applied Anthropology: Essays in Honor of David W. Brokensha.* Boulder, CO: Westview Press.

Chambers, R. 1974. *Managing Rural Development: Ideas and Experience from East Africa.* Uppsala: Scandinavian Institute of African Studies.

———. 1983. *Rural Development: Putting the Last First.* London: Longman.

———. 1997. *Whose Reality Counts? Putting the First Last.* London: Intermediate Technology Publications.

Chandrasekhar, S. 1990. *Third World Development Experience: Tanzania.* Delhi: Daya Publishing House.

Chittick, Neville. 1974. "The coast before the arrival of the Portuguese." In B. A. Ogot, ed., *Zamani: A Survey of East African History.* Nairobi: East African Publishing House.

CIMMYT. 1988. *From Agronomic Data to Farmer Recommendations.* CIMMYT Economics Program. Mexico City: CIMMYT.

Claeson, Claes-Fredrik. 1977. "Interregional population movement and cumulative demographic disparity." *Geografiska Annaler* 56, ser. B:105–120.

Clark, W. C., and N. M. Dickson. 2003. "Sustainability science: The emerging research program." *Proceedings of the National Academy of Science* 100:8059–8061.

Cliffe, Lionel, Peter Lawrence, William Luttrell, Shem Migot-Adholla, and John
    S. Saul, eds. 1975. *Rural Cooperation in Tanzania*. Dar es Salaam: Tanzania
    Publishing House.
Cliffe, Lionel, and John S. Saul, eds. 1972. *Socialism in Tanzania*. 2 vols. Dar es
    Salaam: East African Publishing House.
Cochemé, Jacques, and P. Franquin. 1967. *A Study of the Agroclimatology of the
    Semiarid Area South of the Sahara in West Africa*. FAO/UNESCO/WMO Inter-
    agency Project on Agroclimatology, Technical Reports. Rome: Food and Agri-
    culture Organization of the United Nations.
Coleman, D. C. 1996. *Fundamentals of Soil Ecology*. San Diego, CA: Academic
    Press.
Collier, Paul, Samir Radwan, and Samuel Wangwe. 1986. *Labour and Poverty in
    Rural Tanzania: Ujamaa and Rural Development in the United Republic of Tan-
    zania*. Oxford: Clarendon Press.
Conklin, H. 1957. *Hanunóo Agriculture*. Rome: Food and Agriculture Organiza-
    tion.
Conyers, D., F. Kassulamemba, A. Mbwana, F. Mosha, and P. Nnko. 1970. *Agro-
    economic Zones in North-eastern Tanzania*. Research Report no. 13. Bureau of
    Resource Assessment and Land Use Planning, University of Dar es Salaam,
    Tanzania.
Corbett, John D. 1990. "Agricultural potential from an agro-climate analysis for a
    semiarid area of Kitui District, Kenya." PhD diss., University of Minnesota.
Coulson, Andrew, ed. 1979. *African Socialism in Practice: The Tanzanian Experi-
    ence*. Nottingham, UK: Spokesman.
Dagg, M. 1965. "A rational approach to the selection of crops for areas of marginal
    rainfall in East Africa." *East African Agricultural and Forestry Journal* 30:296–
    300.
Dagg, M., and C. O. Othieno. 1968. "A study of the water use of tea at the Tea Re-
    search Institute of East Africa using a hydraulic lysimeter." Paper presented at
    Fourth Specialist Meeting on Applied Meteorology, Nairobi, 26–27 November.
Dale, Ivan R., and P. J. Greenway. 1961. *Kenya Trees and Shrubs*. Nairobi:
    Buchanan's Kenya Estates.
Darrah, L. L. 1976. "Altitude and environmental responses in the 1972–73 Eastern
    African Maize Variety Trial." *East African Agricultural and Forestry Journal*
    41:273–288.
Darrah, L. L., and L. H. Penny. 1974. "Altitude and environmental responses of
    entries in the 1970–71 Eastern African Maize Variety Trial." *East African Agri-
    cultural and Forestry Journal* 40:77–88.
Datoo, B. A. 1970. "Rhapta: The location and importance of East Africa's first
    port." *Azania* 5:65–75.
Department of Agriculture, Tanganyika Territory. 1945. *Agriculture in Tanganyika*.
    Pamphlet no. 41. Dar es Salaam: The Government Printer.
de Pauw, E. 1984. *Soils, Physiography and Agroecological Zones of Tanzania (Con-
    sultant's Final Report on the . . .)*, Crop Monitoring and Early Warning System
    Project. GCPS/URT/047/NET. Ministry of Agriculture, Dar es Salaam, or the
    United Nations.

de Souza, A. R., and P. W. Porter. 1974. *The Underdevelopment and Modernization of the Third World.* Resource Paper no. 28. Washington, DC: Association of American Geographers.

Dewan, Hari C. 1983. *Soil Moisture Studies under Rainfed Agriculture in Tanzania.* Rome: Food and Agriculture Organization; Muheza: National Soil Service, TARO Agricultural Research Institute, Mlingano, Ministry of Agriculture, Tanzania.

Diak, G. R., M. C. Anderson, W. L. Bland, K. J. M. Norman, J. M. Mecikalski, and R. M. Aune. 1998. "Agricultural management decision aids driven by real-time satellite data." *Bulletin of the American Meteorological Society* 79:1345–1355.

District Books. For Handeni, Lushoto, Pangani, and Tanga Districts. 3 reels of microfilm. Syracuse University Library, Syracuse, NY.

Doorenbos, J., and A. H. Kassam. 1979. *Yield Response to Water.* FAO Irrigation and Drainage Paper 33. Rome: Food and Agriculture Organization.

Doorenbos, J., and W. O. Pruitt. 1977. *Crop Water Requirements.* FAO Irrigation and Drainage Paper 24. Rome: Food and Agriculture Organization.

Doro, Marion E., ed. 1995. "United Republic of Tanzania." *Africa Contemporary Record* 22:B346–B372.

Douglas, Mary. 1963. *The Lele of Kasai.* London: Oxford University Press.

———. 1985. *Risk Acceptability according to the Social Sciences.* London: Routledge.

———. 1986. *Risk Acceptability.* Berkeley and Los Angeles: University of California Press.

———. 1992. *Risk and Blame: Essays in Cultural Theory.* London: Routledge.

Douglas, Mary, and Marcel Calvez. 1990. "The self as risk taker: A cultural theory of contagion in relations to AIDS." *Sociological Review* 38:445–464.

Douglas, Mary, and Adam Wildavsky. 1982. *Risk and Culture.* Berkeley and Los Angeles: University of California Press.

Dowker, B. D. 1963. "Rainfall reliability and maize yields in Machakos District." *East African Agricultural and Forestry Journal* 29:134–138.

———. 1964. "A note on the reduction in yield of Taboran maize by late planting." *East African Agricultural and Forestry Journal* 30:33–34.

Dunne, T. 1979. "Sediment yield and land use in tropical catchments." *Journal of Hydrology* 42:281–300.

Dupriez, G. L. 1964. *L'évaporation et les besoins en eau des différentes cultures dans la région de Muvasi (Bas-Congo).* Serie scientifique no. 106. Yangambi: Institut National pour l'Etude Agronomique au Congo.

East Africa High Commission. 1961. *Quarterly Economic and Statistical Bulletin,* no. 52. Nairobi: E. A. Printers (Boyd).

East African Statistical Department. 1958. *Tanganyika Population Census, General African Census, August 1957.* Part 2, *Territorial Census Areas.* Dar es Salaam: The Government Printer.

East African Statistical Department, Tanganyika Unit. 1959. *Agricultural Census, 1958.* Dar es Salaam: The Government Printer.

———. 1961. *Census of Large Scale Commercial Farming in Tanganyika, October, 1960.* Nairobi: East African Statistical Department.

Eberhart, S. A., L. H. Penny, and M. N. Harrison. 1973. "Genotype by environment interactions in maize in eastern Africa."*East African Agricultural and Forestry Journal* 39:61–71.

Economic Report on Tanga Region. 1971. Dar es Salaam: Bureau of Resource Assessment and Land Use Planning, University of Dar es Salaam.

Economics and Statistics Division, Tanganyika. 1961. *Statistical Abstract, 1961.* Dar es Salaam: The Government Printer.

EcoSystems Ltd. 1986. *Baseline Survey of Machakos District: 1985 and Land Use Changes in Machakos District: 1981–1985.* Report no. 4 for Machakos Integrated Development Project. Nairobi: EcoSystems Ltd.

Edgerton, Robert B. 1971. *The Individual in Adaptation: A Study of Four East African Peoples.* Berkeley and Los Angeles: University of California Press.

Elfring, Wilhelm. 1988. *Production and Marketing of Fruits in the Lowland Areas of Tanga Region, Republic of Tanzania.* Eschborn: German Agency for Technical Cooperation (GTZ).

Engelhardt, W. 1962. *Survival of the Free.* New York: Putnam.

Escobar, A., D. Rocheleau, and S. Kothari. 2002. "Environmental social movements and the politics of place." *Development* 45:28–36.

Feierman, Steven. 1974. *The Shambaa Kingdom: A History.* Madison: University of Wisconsin Press.

———. 1990. *Peasant Intellectuals: Anthropology and History in Tanzania.* Madison: University of Wisconsin Press.

———. 1993. "Defending the promise of subsistence: Population growth and agriculture in the west Usambara Mountains, 1920–1980." In B. L. Turner II, Goran Hyden, and Robert W. Kates, eds., *Population Growth and Agricultural Change in Africa.* Gainesville: University Press of Florida.

Fernando, J. L. 2003. "Rethinking sustainable development." *Annals of the American Academy of Political and Social Science* 590:1–266.

FEWS NET. 2003. *Tanzania Monthly Report,* 14 November. Famine Early Warning Systems Network. http://www.fews.net/centers/current/monthlies/report.

———. 2004a. *Tanzania Food Security Report,* 17 February. Famine Early Warning Systems Network. http://www.fews.net/centers/current/monthlies/report.

———. 2004b. *Tanzania Food Security Report,* 16 March. Famine Early Warning Systems Network. http://www.fews.net/centers/current/monthlies/report.

Fleuret, Patrick C. 1978. "Farm and market: A study of society and agriculture in Tanzania." PhD diss., University of California, Santa Barbara.

———. 1980. "Sources of inequality in Lushoto District, Tanzania." *African Studies Review* 23:69–87.

Food Studies Group. 1992a. *Agricultural Diversification and Intensification Study, Final Report.* Vol. 1, *Findings and Policy Implications.* Oxford: International Development Centre, University of Oxford; Morogoro: Department of Rural Economy, Sokoine University of Agriculture.

———. 1992b. *Agricultural Diversification and Intensification Study, Final Report.* Vol. 2, *Farming Systems: Characteristics and Trends.* Oxford: International Development Centre, University of Oxford; Morogoro: Department of Rural Economy, Sokoine University of Agriculture.

———. 1992c. *Agricultural Diversification and Intensification Study, Final Report.* Vol. 3, *Summary of Key Results.* Oxford: International Development Centre, University of Oxford; Morogoro: Department of Rural Economy, Sokoine University of Agriculture.

Ford, John. 1971.*The Role of Trypanosomiases in African Ecology: A Study of the Tsetse Fly Problem.* Oxford: Clarendon Press.

Forest Division. 1984. "Tanzania, Vegetation Cover Types." Map, scale: 1:2,000,000, based on *Landsat* satellite images and field observation. Dar es Salaam: Ministry of Lands, Natural Resources, and Tourism.

Forsyth, T. 2003. *Critical Political Ecology: The Politics of Environmental Science.* London: Routledge.

Frake, C. O. 1962. "Cultural ecology and ethnography." *American Anthropologist* 62:53–59.

Freire, P. 1970. *Pedagogy of the Oppressed.* New York: Seabury Press.

Frere, M., and L. I. Popov. 1979. *Agrometeorological Crop Monitoring and Forecasting.* FAO Production and Protection Paper 17. Rome. (This report formed the basis on which the Famine Early Warning System for Tanzania was established and which continues to be run out of TANCOT House, Dar es Salaam.)

Freyhold, Michaela von. 1971. *Case Studies from Tanga Coast.* 2 vols. Dar es Salaam: University of Dar es Salaam.

———. 1972. *The Potential for* Ujamaa *in the Coastal and Semi-coastal Areas of Tanga Region.* BRALUP Seminar Paper. Dar es Salaam: University of Dar es Salaam.

———. 1977. *Universal Primary Education and Education for Self-Reliance in Tanga: An Evaluation of the TIRDEP School Building Programme and an Outline of Future Tasks in the Field of Primary and Poly-technical Education.* Dar es Salaam: University of Dar es Salaam.

———. 1979. Ujamaa*Villages in Tanzania: Analysis of a Social Experiment.* New York: Monthly Review Press.

Fuggles-Couchman, N. R. 1964. *Agricultural Change in Tanganyika, 1945–1960.* Stanford: Food Research Institute.

Geiger, Susan. 1982. "*Umoja wa wanawake wa Tanzania* and the needs of the rural poor." *African Studies Review* 25:45–65.

———. 1996. "Tanganyika nationalism as 'Women's Work': Life histories, collective biography and changing historiography." *Journal of African History* 37:465–478.

Gerlach, Luther P. 1965. "Nutrition in its sociocultural matrix: Food getting and using along the East African coast." In David Brokensha, ed., *Ecology and Economic Development in Tropical Africa,* Berkeley and Los Angeles: University of California Press.

German Agency for Technical Cooperation. 1975. *The Present Central Places Pattern of Tanga Region.* Technical Report no. 7, *Tanga Water Master Plan, Tanga Region.* Essen: Agrar- und Hydrotechnik.

———. 1976a. *Tanga Water Master Plan, Tanga Region.* Vol. 3, *Groundwater Resources.* Essen: Agrar- und Hydrotechnik.

———. 1976b. *Tanga Water Master Plan, Tanga Region.* Vol. 5, *Socio-economics.* Essen: Agrar- und Hydrotechnik.

———. 1976c. *Tanga Water Master Plan, Tanga Region.* Vol. 6, *Agriculture.* Essen: Agrar- und Hydrotechnik.

German Development Service. 2003. "Community involvement in forest management, Handeni District." *Community Based Forest Management, Handeni.* http://www.ded-tanzania/de/fglw/cbfh.html.

Giblin, James L. 1990. "Trypanosomiasis control in African history: An evaded issue?" *Journal of African History* 31:59–80.

———. 1992. *The Politics of Environmental Control in Northeastern Tanzania, 1840–1940.* Philadelphia: University of Pennsylvania Press.

Gichuki, F. D. 1991. *Environmental Change and Dryland Management in Machakos District, Kenya, 1930–90, Conservation Profile.* Working Paper 56. London: Overseas Development Institute.

Gillman, Clement. 1942. "A short history of Tanganyika railways." *Tanganyika Notes and Records* 13:14–56.

Glaeser, Bernhard. 1984. *Ecodevelopment in Tanzania: An Empirical Contribution on Needs, Self-Sufficiency, and Environmentally-Sound Agriculture on Peasant Farms.* Berlin: Mouton.

Glassman, Jonathon. 1991. "The bondsman's new clothes: The contradictory consciousness of slave resistance on the Swahili coast." *Journal of African History* 32:277–312.

Goldschmidt, Walter R. 1976. *Culture and Behavior of the Sebei.* Berkeley and Los Angeles: University of California Press.

Goldson, J. R. 1963. "The effect of time of planting on maize yields." *East African Agricultural and Forestry Journal* 29:160–163.

Gommes, R. 1985. "The Tanzanian Crop Monitoring and Early Warning Systems Project." Mimeographed.

Gommes, R. A., and M. Houssian. 1982. *Rainfall Variability, Types of Growing Seasons and Cereal Yields in Tanzania Crop Monitoring and Early Warning Systems Project.* Dar es Salaam: FAO/Kilimo.

Grossman, L. S. 1977. "Man-environment relationships in anthropology and geography." *Annals of the Association of American Geographers* 67:126–144.

Guillain, Charles. 1843. "Côte d'Ivoire de Zanguebar et Mascate, 1841." *Revue Coloniale,* pp. 520–563.

Gupta, S. C., and W. E. Larson. 1982. "Modeling soil mechanical behavior during tillage." In P. W. Unger et al., eds., *Symposium on Predicting Tillage Effects on Soil Physical Properties and Processes,* 151–178. Special Publication 44. Madison, WI: American Society of Agronomy.

Gwassa, G. C. K. 1969. "The German intervention and African resistance in Tanzania." In I. N. Kimambo and A. J. Temu, eds., *A History of Tanzania.* Nairobi: East African Publishing House.

Hamilton, A. C. 1989. "Distribution of tree species in the East Usambara Mountains." In A. C. Hamilton and R. Bensted-Smith, eds., *Forest Conservation in the East Usambara Mountains, Tanzania.* Gland, Switzerland: IUCN The World Conservation Union.

Hamilton, A. C., and R. Bensted-Smith, eds. 1989. *Forest Conservation in the East Usambara Mountains, Tanzania.* Gland, Switzerland: IUCN The World Conservation Union.

Hartemink, Alfred E. 1995. *Soil Fertility Decline under Sisal Cultivation in Tanzania.* Wageningen: Netherlands International Soil Reference and Information Centre.

Hathout, Salah, and S. Sumra. 1974. *Rainfall and Soil Suitability Index for Maize Cropping in Handeni District.* Dar es Salaam: Bureau of Resource Assessment and Land Use Planning, University of Dar es Salaam.

Heijnen, J., and R. W. Kates. 1974. "Northeast Tanzania: Comparative observations along a moisture gradient." In Gilbert F. White, ed., *Natural Hazards: Local, National, Global.* New York: Oxford University Press.

Heijnen, J., and J. Kreysler. 1971. "Cooperative vegetable production schemes in Lushoto District." Paper presented at the East African Agricultural Economics Society Conference, Nairobi, 24–27 June.

Hewitt, K. 1987. "The social space of terror: Towards a civil interpretation of total war." *Environment and Planning D: Society and Sp*ace 5:445–474.

Hill, M. F. 1962. *Permanent Way: The Story of the Tanganyika Railways,* vol. 2. Nairobi: East African Railways and Harbours.

Hoyle, B. S. 1987. *Gillman of Tanganyika, 1882–1946: The Life and Work of a Pioneer Geographer.* Aldershot, UK: Brookfield.

Hubert, K., and F. Sander. 1978. *The Rehabilitation of Rural Roads in Handeni District (Tanzania).* Eschborn: German Agency for Technical Cooperation (GTZ).

Humphries, S., J. Gonzales, J. Jiminez, and F. Sierra. 2000. *Searching for Sustainable Land Use Practices in Honduras.* Network Paper no. 104. London: Overseas Development Institute. http://www.pnas.org/cgi/content/full/100/14/8086.

Iliffe, John. 1969. *Tanganyika under German Rule, 1905–1912.* Cambridge: Cambridge University Press.

———. 1971. *Agricultural Change in Modern Tanganyika.* Historical Association of Tanzania, Paper 10. Nairobi: East African Publishing House.

———. 1974. "Tanzania under German and British rule." In B. A. Ogot, ed., *Zamani: A Survey of East African History.* Nairobi: East African Publishing House.

———. 1979. *A Modern History of Tanganyika.* African Studies Series 25. Cambridge: Cambridge University Press.

Inter-academy Council. 2004. *Realizing the Promise and Potential of African Agriculture: Science and Technology Strategies for Improving Agricultural Productivity and Food Security in Africa.* Report launched at United Nations Headquarters, 25 June 2004. Amsterdam: Inter-academy Council, c/o Royal Netherlands Academy of Arts and Sciences. http://www.interacademycouncil.net.

International Bank for Reconstruction and Development. 1960. *The Economic Development of Tanganyika.* Dar es Salaam: The Government Printer.

International Institute of Tropical Agriculture. 1999. *Annual Report.* Ibadan, Nigeria.

ISNAR. 1996. *Ecoregional Fund to Support Methodological Initiatives.* The Hague, Netherlands: International Service for National Agricultural Research.

Iversen, Svein Terje. 1991. *The Usambara Mountains, NE Tanzania: Phytogeography of the Vascular Plant Flora*. Acta Universitis Upsaliensis, Symbolae Botanicae Upsaliensis 29, no. 3. Uppsala: Uppsala University.

Johnson, Frederick. 1939. *A Standard Swahili-English Dictionary*. Oxford: Oxford University Press.

Jones, C. A., and J. R. Kiniry, eds. 1986. *CERES-Maize: A Simulation Model of Maize Growth and Development*. College Station: Texas A&M University Press.

Kabat, P., B. J. van den Broek, and R. A. Feddes. 1992. "SWACROP: A water management and crop production simulation model." *ICID Bulletin* 41:61–84.

Kahama, C. George, T. L. Maliyamkono, and Stuart Wells. 1986. *The Challenge for Tanzania's Economy*. London: Heinemann.

Kaihura, Fidelis B. S. 1991.*Soil Erosion and Conservation Studies in Tanzania, Opportunities and Strategies for Research*. Soil Fertility and Management Report F8, Mlingano Agricultural Research Institute, National Soil Service, Ministry of Agriculture and Livestock Development, Tanga.

Kaluli, J. W. 1992. "NGOs and technological change." In M. Tiffen, ed., *Environmental Change and Dryland Management in Machakos District, Kenya, 1930–90, Institutional Profile*. Working Paper. London: Overseas Development Institute.

Kasperson, J. X., and R. E. Kasperson, eds. 2001. *Global Environmental Risk*. Tokyo: United Nations University Press.

Kasperson, J. X., R. E. Kasperson, and B. L. Turner II, eds. 1995. *Regions at Risk: Comparisons of Threatened Environments*. Tokyo: United Nations University Press.

Kates, R. W. 1995. "Labnotes from the Jeremiah experiment: Hope for a sustainable transition." *Annals of the Association of American Geographers* 85:632–640.

———. 2000. "Population and consumption: What we know, what we need to know." *Environment* 42:10–19.

Kates, R. W., W. C. Clark, R. Correll, J. M. Hall, C. C. Jaeger, I. Lowe, J. J. McCarthy, H. J. Schellnbuber, B. Bolin, N. M. Dickson, et al. 2001. "Sustainability science." *Science* 292:641–642.

Kates, R. W., J. McKay, and L. Berry. 1970. "Twelve new settlements in Tanzania: A comparative study of success." In *Geography Papers*. Dar es Salaam: University of Dar es Salaam.

Kaya, Hassan Omari. 1985. *Problems of Regional Development in Tanzania: A Case Study of the Tanga Region*. Saarbrücken and Fort Lauderdale: Breitenbach.

Kayamba, H. M. T. 1947. "Notes on the Wadigo." *Tanganyika Notes and Records* 23:80–96.

Kikula, Idris S. 1989. "Spatial changes in forest cover on the East Usambara Mountains." In A. C. Hamilton and R. Bensted-Smith, eds., *Forest Conservation in the East Usambara Mountains, Tanzania*. Gland, Switzerland: IUCN The World Conservation Union.

Kimambo, I. N. 1996. "Environmental control and hunger in the mountains and plains of nineteenth-century northeastern Tanzania." In G. Maddox, J. L. Giblin, and I. N. Kimambo, eds., *Custodians of the Land: Ecology and Culture in the History of Tanzania*. London: James Currey.

Kimambo, I. N., and A. J. Temu. 1969. *A History of Tanzania.* Nairobi: East African Publishing House.

Kiondo, E. 1999. "Access to gender and development information by rural women in Tanzania." *Innovation* 19: not paginated, http://www.library.unp.ac.za/innovation/id45.htm.

Kirda, C., ed. 1999. *Crop Yield Response to Deficit Irrigation.* Report of FAO/IAEA Coordinated Research Program by Using Nuclear Techniques: Executed by the Soil and Water Management and Crop Nutrition Section of the Joint FAO/IAEA Division of Nuclear Techniques in Food and Agriculture. Dordrecht: Kluwer Academic Publishers.

Kirio, A., and C. Juma, eds. 1989. *Gaining Ground.* Nairobi: Africa Centre for Technology Study.

Kiro, Selemani. 1953. "The history of the Zigua tribe." *Tanganyika Notes and Records* 34:70–74.

Kjekshus, Helge. 1977. *Ecology Control and Economic Development in East African History: The Case of Tanganyika, 1850–1950.* London: Heinemann.

Koponen, Juhani. 1994. *Development for Exploitation: German Colonial Policies in Mainland Tanzania, 1884–1914.* Helsinki: Tiedekirja; Hamburg: Lit Verlag.

Kreysler J., and M. Mndeme. 1975. "The nutritional status of preschool village children in Tanzania: Observations in Lushoto District." *Ecology and Food Nutrition* 4:15–26.

Kuczynski, R. R. 1949. *Demographic Survey of the British Colonial Empire,* vol. 2. London: Oxford University Press.

Kurtz, Laura L. 1978. *Historical Dictionary of Tanzania.* African Historical Dictionaries, no. 15. Metuchen, NJ, and London: Scarecrow Press.

Kustas, W. P., and J. M. Norman. 1999. "Evaluation of soil and vegetation heat flux predictions using a simple two-source model with radiometric temperatures for partial canopy cover." *Agriculture Forestry Meteorology* 94:13–29.

Lambrecht, Frank L. 1964. "Aspects of evolution and ecology of tsetse flics and trypanosomiasis in prehistoric African environment." *Journal of African History* 5:1–24.

Legum, Colin, ed. 1977. "United Republic of Tanzania." *Africa Contemporary Record* 9:B339–B371.

———, ed. 1981/82. "United Republic of Tanzania." *Africa Contemporary Record* 14:B272–B297.

———, ed. 1988. "United Republic of Tanzania." *Africa Contemporary Record* 19: B419–B458.

———, ed. 1989. "United Republic of Tanzania." *Africa Contemporary Record* 20: B406–B436.

———, ed. 1998. "United Republic of Tanzania." *Africa Contemporary Record* 23: B389–B404.

———, ed. 2000. "United Republic of Tanzania." *Africa Contemporary Record* 24: B401–B421.

Listowel, Judith. 1965. *The Making of Tanganyika.* London: Chatto and Windus.

Little, Peter D., and Michael M Horowitz, with A. Endre Nyerges. 1987. *Lands at Risk in the Third World: Local-Level Perspectives.* IDA Monographs in Development Anthropology. Boulder, CO: Westview Press.

Lofchie, Michael F. 1965. *Zanzibar: Background to Revolution.* Princeton: Princeton University Press.

———. 1989. *The Policy Factor: Agricultural Performance in Kenya and Tanzania.* Nairobi: Heinemann Kenya.

Mabogunje, A. L., and R. W. Kates. 2004. "Sustainable development in Ijebu-Ode, Nigeria: The role of social capital, participation, and science and technology." Center for International Development Working Paper no. 102, www.cid.harvard.edu/cidwp/102.htm.

Machakos District Gazetteer. 1958. Machakos: District Agricultural Office.

Mackenzie, A. F. D. 1998. *Land, Ecology and Resistance in Kenya, 1880–1952.* Edinburgh: Edinburgh University Press for the International African Institute.

Maddox, Gregory, James L. Giblin, and Isaria N. Kimambo. 1996. *Custodians of the Land: Ecology and Culture in the History of Tanzania.* London: James Currey.

Maher, C. 1937. *Soil Erosion and Land Utilization in the Ukamba Reserve: A Report for Colony of Kenya.* Rhodes House Library, Oxford, Afr.S. 755. Nairobi: Government Printer.

Maliyamkono, T. L., and M. S. D. Bagachwa. 1989. *The Second Economy in Tanzania.* London: James Currey.

Manning, H. L. 1956. "The statistical assessment of rainfall probability and its application in Uganda agriculture." *Research Memoirs* (Empire Cotton Growing Corp.) 23:460–480.

Maro, P. S. 1974. "Population and land resources in northern Tanzania: The dynamics of change, 1920–1970." PhD diss., University of Minnesota, Minneapolis.

Mascarenhas, Adolfo C. 1970. "Resistance and change in the sisal plantation system of Tanzania." PhD diss., University of California, Los Angeles.

———. 1971a. "Agricultural vermin in Tanzania." In S. H. Ominde, ed., *Studies in East African Geography and Development.* London: Heinemann.

———. 1971b. "Sisal." In Len Berry, ed., *Tanzania in Maps.* London: University of London Press.

———. 1973. "An introduction to Tanzanian studies on famine and malnutrition." In Adolfo C. Mascarenhas, ed., "Studies in famines and food shortages," special issue, *Journal of the Geographical Association of Tanzania* 8:1–12.

———. 1975. *Drought and the Optimization of Tanzania's Environmental Potential.* Stockholm: Swedish Institute of Ecological Studies.

———. 2000. "Poverty, environment and livelihood along the highland–lowland gradient of the Usambaras in Tanzania." Book-length manuscript. (Part of this study has been published as "Poverty, environment and livelihood along the highland–lowland gradient of the Usambaras in Tanzania," *REPOA* Publication* 00.2, Dar es Salaam, 57 pp.; and the second booklet, titled "Poverty, livelihood and food insecurity along the gradients of the Usambaras," Dar es Salaam, 67 pp., has been submitted for publication.)

_____. n.d. [c. 1972.] "Famine and food shortages in Tanganyika (1925–1945)."
Draft version of paper prepared for *Journal of the Geographical Association of
Tanzania*. Typescript.

Mbilinyi, Marjorie. 1971. *The Participation of Women in African Economies.* Economic Research Bureau Paper 71.12. Dar es Salaam: University of Dar es
Salaam.

Mbithi, Philip M. 1977. "Human factors in agricultural management in East Africa."
Food Policy 2(1):27–33.

McBride, R. A., and E. E. Mackintosh. 1984. "Soil survey interpretations from water
retention data: I. Development and validation of a water retention model." *Soil
Science Society of America Journal* 48:1338–1343.

McCulloch, J. S. G. 1965. "Tables for the rapid computation of the Penman estimate of evaporation." *East African Agricultural and Forestry Journal* 30:286–
295.

McKay, J., A. Daraja, and W. Mlay. 1972. *Interim Report on a Base-line Study of
the Proposed Village Settlement at Kiwanda.* Research Report no. 3, Bureau of
Resource Assessment and Land Use Planning. Dar es Salaam: University of
Dar es Salaam.

Merten, Peter. 1989. *Research of the Self-Help Potential of Villagers in Handeni
District.* Tanga: Regional Community Development Office; Eschborn: Village
Development Programme Tanga Region, German Agency for Technical Coop-
eration.

Mhando, Helen P. 1975. "Capitalist penetration and the growth of peasant agriculture in Korogwe District, 1920–1975." MA thesis, University of Dar es Salaam.

Mhita, Mohamed Salim. 1984. *The Use of Water Balance Models in the Optimization
of Cereal Yields in Seasonally-Arid Tropical Regions.* Reading: Department of
Meteorology, University of Reading.

Mhita, Mohamed Salim, and I. R. Nassib. 1987. *The Onset and End of Rains in
Tanzania.* Dar es Salaam: Directorate of Meteorology.

Michaud, N., K. P. B. Thomson, J. Boisvert, and A. Viau. 1999. *Extraction of
Crop Information Using Multispectral RADARSAT Imagery and Integration
in the SWAP Crop Growth Model.* Quebec City: Université Laval. http://www.
estec.esa.nl/conferences/98C07/papers/P035.pdf.

Middleton, John. 1961. *Land Tenure in Zanzibar.* London: H.M. Stationery Office.

Ministry of Agriculture and Cooperative Development. 1961. *Annual Report of the
Department of Agriculture, 1960.* Dar es Salaam: The Government Printer.

Mkama, J. 1969. *Transport Planning in Tanzania: An Assessment.* Research Paper
no. 8, Bureau of Resource Assessment and Land Use Planning. Dar es Salaam:
University of Dar es Salaam.

Molina, J. A. E., and P. Smith. 1997. "Modeling carbon and nitrogen processes in
soils." *Advances in Agronomy* 62:253–298.

Monteith, J. L. 1965. "Light distribution and photosynthesis in field crops." *Annals
of Botany,* n.s., 29:17–37.

_____. 1972. "Solar radiation and productivity in tropical ecosystems." *Journal of
Applied Ecology* 9:747–766.

Moock, J. L., and R. E. Rhoades, eds. 1992. *Diversity, Farmer Knowledge, and Sustainability: Food Systems and Agrarian Change.* Ithaca, NY: Cornell University Press.

Mörth, H. T. 1971. "On the distribution of rain showers in the Tanga Region of Tanzania." In Gerhard Tshcannerl, ed., *Water Supply.* Research Paper no. 20, Bureau of Resource Assessment and Land Use Planning. Dar es Salaam: University of Dar es Salaam.

Mortimore, Michael. 1998. *Roots in the African Dust: Sustaining the Sub-Saharan Drylands.* Cambridge: Cambridge University Press.

———. 1999. *Adapting to Drought: Farmers, Famines and Desertification in West Africa.* Cambridge: Cambridge University Press.

Mortimore, Michael, and W. M. Adams. 1999. *Working the Sahel: Environment and Society in Northern Nigeria.* London: Routledge.

Mortimore, Michael, and Mary Tiffen. 1994. "Population growth and a sustainable environment." *Environment* 36:10–32.

Mortimore, Michael, and K. Wellard. 1991. *Environmental Change and Dryland Management in Machakos District, Kenya, 1930–90, Profile of Technological Change.* Working Paper 57. London: Overseas Development Institute.

Moseley, W. G., and B. I. Logan, eds. 2004. *African Environment and Development: Rhetoric, Programs, Realities.* Aldershot, UK: Ashgate Publishing.

Mpigachapa wa Serikali. 1990. *Hali ya Uchumi wa Taifa katika Mwaka 1989.* Dar es Salaam: Mpigachapa wa Serikali.

Mshana, A. M. E., and R. S. S. Mduma. 2000. "Animal power for weed control: Experiences of MATI Mlingano, Tanga, Tanzania." In P. Starkey and T. Simalenga, eds., *Animal Power for Weed Control.* Wageningen, Netherlands: Technical Centre for Agricultural and Rural Cooperation.

Mukandala, Rwekaza S., and Haroub Othman. 1994. *Liberalization and Politics: The 1990 Elections in Tanzania.* Dar es Salaam: Dar es Salaam University Press.

Müller, A. 2003. "A flower in full blossom? Ecological economics at the crossroads between normal and post-normal science." *Ecological Economics* 45:19–27.

Munro, J. Forbes. 1983. "British rubber companies in East Africa before the First World War." *Journal of African History* 24:369–379.

Murton, J. 1999. "Population growth and poverty in Machakos, Kenya." *Geographical Journal* 165:37–46.

National Research Council. 1982. *Ecological Aspects of Development in the Humid Tropics.* Washington, DC: National Academy Press.

National Soil Service. 1988a. "Progress report for 1987/88." Paper presented at the Annual Soils and Fertilizer Use Coordinating Committee Meeting, Arusha, Tanzania, 14–16 November.

———. 1988b. *Soils of Pongwe Estate and Their Potential for Hybrid Sisal Cultivation.* Detailed Soil Survey Report D15. Tanga: Mlingano Agricultural Research Institute, National Soil Service, Ministry of Agriculture and Livestock Development.

———. 1989a. "Progress report for 1988/89." Paper presented at the Annual Soils and Fertilizer Use Coordinating Committee Meeting, Dar es Salaam, 5–6 October.

————. 1989b. *Towards Sustainable Land Use in the East Usambara Mountains.* Site Evaluation Report S12. Tanga: TARO—Mlingano Agricultural Research Institute, National Soil Service.

————. 1990. "Progress report for 1989/90." Paper presented at the Annual Soils and Fertilizer Use Coordinating Committee Meeting, Arusha, Tanzania, 27–28 November.

————. 1992a. *List of Publications, August 1992.* Tanga: Mlingano Agricultural Research Institute, National Soil Service, Ministry of Agriculture and Livestock Development.

————. 1992b. "Progress report for 1991/92." Paper presented at the Annual Soils and Fertilizer Use Coordinating Committee Meeting, Tanga, 14–15 December.

National Steering Committee for NGO Policy. 2000. *The National Policy on Non-governmental Organizations (NGOs) in Tanzania.* Dar es Salaam: United Republic of Tanzania.

Neumann, R. P. 2001. "Disciplining peasants in Tanzania: From state violence to self-surveillance in wildlife conservation." In N. L. Peluso and M. Watts, eds., *Violent Environments.* Ithaca, NY: Cornell University Press.

Ngatunga, Edward L. N. 1981. "Soil erosion studies at Mlingano on the eastern Usambara Uplands." MS thesis, University of Dar es Salaam.

Ngware, S. S. 1982. "Sociological study: Soil erosion control agroforestry project and livestock development in the western Usambaras." Paper prepared for TIRDEP, University of Dar es Salaam, Institute of Development Studies.

Nieuvolt, S. 1973. *Rainfall and Evaporation in Tanzania.* Research Paper no. 24, Bureau of Resource Assessment and Land Use Planning. Dar es Salaam: University of Dar es Salaam.

Nimtz, August H., Jr. 1980. *Islam and Politics in East Africa: The Sufi Order in Tanzania.* Minneapolis: University of Minnesota Press.

Norgaard, Richard B. 1994. *Development Betrayed: The End of Progress and a Coevolutionary Revisioning of the Future.* London: Routledge.

Nyerere, Julius K. 1966. *Freedom and Unity: Uhuru na Umoja.* Dar es Salaam: Oxford University Press.

————. 1967. *Socialism and Rural Development.* Dar es Salaam: The Government Printer.

Ogot, B. A., ed. 1974. *Zamani: A Survey of East African History.* Nairobi: East African Publishing House.

Oliver, Roland, and Gervase Mathew. 1963. *History of East Africa.* London: Oxford University Press.

Parris, T. M., and R. W. Kates. 2003. "Characterizing a sustainability transition: Goals, targets, trends, and driving forces." *Proceedings of the National Academy of Science* 100:8068–8073, http://www.pnas.org/cgi/content/full/100/...& firstpage=8068&resourcetype=1.

Parsons, T. H. 1999. *The African Rank-and-File: Social Implications of Colonial Service in the King's African Rifles, 1902–1964.* Portsmouth, NH: Heinemann.

Pearce, D. G., E. Barbier, and A. Markandya. 1990. *Sustainable Development: Economics and Environment in the Third World.* Aldershot, UK: Edward Elgar.

Peet, R., and M. Watts. 1996. *Liberation Ecologies: Environment, Development, Social Movements*. London: Routledge.

Peluso, N. L., and M. Watts, eds. 2001. *Violent Environments*. Ithaca, NY: Cornell University Press.

Penman, H. L. 1948. "Natural evaporation from open water, bare soil and grass." *Proceedings of the Royal Society of London (A)* 193:120–145.

———. 1963. *Vegetation and Hydrology*. Technical Communication no. 53. Harpenden, UK: Commonwealth Bureau of Soils.

Pereira, L., et al., eds. 1995. "Workshop of crop-water-models." In *Crop-Water Simulation Models in Practice: Selected Papers of the 2nd Workshop in Crop-Water-Models Held on the Occasion of the 15th Congress of the International Commission on Irrigation and Drainage (ICID) at The Hague, the Netherlands in 1993*. Wageningen: Wageningen Pers.

Pike, K. L. 1966. "Etic and emic standpoints for the description of behavior." In A. G. Smith, ed., *Communication and Culture*. New York: Holt, Rinehart, and Winston.

Pilbeam, C. J., C. C. Daamen, and L. P. Simmonds. 1995. "Analysis of water budgets in semi-arid lands from soil water records." *Experimental Agriculture* 31:131–149.

Porter, Philip W. 1965. "Environmental potentials and economic opportunities—a background for cultural adaptation." *American Anthropologist* 67:409–420.

———. 1976. "Agricultural development and agricultural vermin." Paper presented at the annual meeting of the American Academy for the Advancement of Science, Boston, 22 February.

———. 1978. "Geography as human ecology: A decade of progress in a quarter century." *American Behavioral Science* 22:15–39.

———. 1979. *Food and Development in the Semi-arid Zone of East Africa*. Foreign and Comparative Studies/African Series 32. Syracuse: Syracuse University.

———. 1981. "Problems of agro-meteorological modeling in Kenya." *Interciencia* 6 (July/August):226–233. (Summary in Spanish, 6:297–298.)

———. 1983. "Problems of agrometeorological modeling in Kenya." In David F. Cusack, ed., *Agroclimate Information for Development: Reviving the Green Revolution*. Boulder, CO: Westview Press.

———. 2001. "Ecology, cultural." In Neil J. Smelser and Paul B. Baltes, eds., *International Encyclopedia of the Social and Behavioral Sciences*, 6:4035–4041.

Porter, Philip W., and Gregory M. Flay. 1998. "Materials for the historical geography of Tanzanian agriculture: Some maps from adjectives." *East African Geographical Review* 20:39–57.

Porter, Philip W., and Patricia G. Porter. 1993. "Eight months in Tanzania." Unpublished journal kept during fieldwork, 1992–1993, in Tanga Region, Tanzania.

Porter, Philip W., and Eric S. Sheppard. 1998. *A World of Difference: Society, Nature, Development*. New York: Guilford Press.

Potts, W. H. 1937. "The distribution of tsetse flies in Tanganyika Territory." *Bulletin of Entomological Research* 28:129–148.

Raikes, P. L. 1972. *Village Planning for Ujamaa*. Dar es Salaam: Economic Research Bureau, University of Dar es Salaam.

Ratcliff, L. F., J. T. Ritchie, and D. K. Cassel. 1983. "Field-measured limits of soil water availability as related to laboratory-measured properties." *Soil Science Society of America* 47(4):770–775.

Rawls, J. 1971. *A Theory of Justice.* Cambridge: Harvard University Press.

Rawls, Walter J. 1982. "Estimating soil water retention from soil properties." *Journal of the Irrigation and Drainage Division, Proceedings of the American Society of Civil Engineers* 108(IR2):166–171.

Regional Planning Office, Tanga. 1983. *Muheza District Development Strategy.*

Reiches-Kolonialamt. 1914. *Die deutschen Schutzgebiete in Afrika und der Südsee, 1912/13.* Berlin: Reiches-Kolonialamt.

Richards, P. 1985. *Indigenous Agricultural Revolution: Ecology and Food Production in West Africa.* Boulder, CO: Westview Press.

————. 1986. *Coping with Hunger: Hazard and Experiment in an African Rice-Farming System.* London: Allen and Unwin.

Rocheleau, D. E. 1991. "Gender, ecology, and the science of survival: Stories and lessons from Kenya." *Agriculture and Human Values* 8:156–165.

Rocheleau, D. E., and D. Edmunds. 1997. "Women, men and trees: Gender, power and property in forest and agrarian landscapes." *World Development* 25:1351–1371.

Rocheleau, D. E., P. Benjamin, and A. Diang'a. 1995. "The Ukambani region of Kenya." In J. E. Kasperson, R. E. Kasperson, and B. L. Turner II, eds., *Regions at Risk: Comparisons of Threatened Environments.* Tokyo: United Nations Press.

Rocheleau, D. E., P. Steinberg, and P. Benjamin. 1995. "Environment, development, crisis and crusade: Ukambani, Kenya, 1890–1990." *World Development* 23:1037–1051.

Rostom, R. S., and M. Mortimore. 1991. *Environmental Change and Dryland Management in Machakos District, Kenya, 1930–90, Land Use Profile.* Working Paper 58. London: Overseas Development Institute.

Rubin, Deborah S. 1985. "People of good heart: Rural response to economic crisis in Tanzania." PhD diss., Johns Hopkins University.

Ruffo, C. K. 1989. "Some useful plants of the eastern Usambaras." In A. C. Hamilton and R. Bensted-Smith, *Forest Conservation in the East Usambara Mountains, Tanzania.* Gland, Switzerland: IUCN The World Conservation Union.

Ruffo, C. K., I. V. Mwasha, and C. Mmari. 1989. "The use of medicinal plants in the East Usambaras." In A. C. Hamilton and R. Bensted-Smith, *Forest Conservation in the East Usambara Mountains, Tanzania.* Gland, Switzerland: IUCN The World Conservation Union.

Ruttan, V. W. 1994. "Constraints on the design of sustainable systems of agricultural production." *Ecological Economics* 10:209–219.

Sadleir, Randal. 1999. *Tanzania: Journey to Republic.* London: Radcliffe Press.

Samatar, A., and A. I. Samatar. 1987. "The material roots of the suspended African state: Arguments from Somalia." *Journal of Modern African Studies* 25:669–690.

Sarris, Alexander H., and Rogier van den Brink. 1991. *Economic Policy and Household Welfare during Crisis and Adjustment in Tanzania.* Washington, DC: Cornell Food and Nutrition Policy Program.

Sayer, Gerald F., ed. 1930. *The Handbook of Tanganyika*. London: Macmillan and Co.

Scheinman, David. 1986. *Caring for the Land of the Usambaras*. Report prepared for TIRDEP-Soil Erosion Control and Agroforestry Project (SECAP), Lushoto, Tanzania.

————. 1988. "Women's Participation Seminar." Mbeya, March. Typescript.

————. 1989. *The Feasibility of Tractor Mechanization in Tanga Region*. Tanga: Tanga Integrated Rural Development Program.

Schnee, Heinrich. 1920. *Deutsches Kolonial-Lexikon*. Leipzig: Verlag von Quelle und Meyer.

Schwarz, Michiel, and Michael Thompson. 1990. *Divided We Stand: Redefining Politics, Technology and Social Choice*. Philadelphia: University of Pennsylvania Press.

SCOPE. 1978. *Report on the Workshop on Climate/Society Interface*. Paris: Scientific Committee on Problems of the Environment.

Seppälä, Pekka. 1998. *Diversification and Accumulation in Rural Tanzania: Anthropological Perspectives on Village Economics*. Stockholm: Nordiska Afrikainstitutet.

Smith, M., E. Weltzein, L. Meitzner, and L. Sperling. 2004. *Technical and Institutional Issues in Participatory Plant Breeding from the Perspective of Formal Plant Breeding: A Global Analysis of Issues, Results, and Current Experience*. Working Document 3, PRGA Program, Cali, Colombia. http://www.ciat.org/ipra/ing/index.htm.

Sosovele, H. 1999. "The challenges of animal traction in Tanzania." In P. Starkey and P. Kaumbutho, eds., *Meeting the Challenges of Animal Traction*. Harare, Zimbabwe: Intermediate Technology Publications.

Spear, T. 1996. "Struggles for the land: The political and moral economies of land on Mount Meru." In G. Maddox, J. Giblin, and I. N. Kimambo, eds., *Custodians of the Land: Ecology and Culture in the History of Tanzania*. London: James Currey.

Spencer, Dunstan. 2001. "Will they survive? Prospects for small farmers in Sub-Saharan Africa." In *Sustainable Food Security for All by 2020*. Washington, DC: International Food Policy Research Institute. http://www.ifpri.org/2020 conference/PDF/summary_spencer.pdf.

Sperling, L., and C. Langley. 2002. "Beyond seeds and tools: Effective support to farmers in emergencies." *Disasters* 26:283–287.

Steward, J. H. 1968. "Cultural ecology." In D. J. Sills, ed., *International Encyclopedia of the Social Sciences*, 4:337–344. New York: Macmillan.

Stewart, J. Ian. 1980. "Effective rainfall analysis to guide farm practices and predict yields." Paper presented at the Fourth Annual General Meeting of the Soil Science Society of East Africa, Nairobi, 26–27 November.

————. 1982. "Crop yields and returns under different soil moisture regimes." Paper presented at the Third FAO/SIDA Seminar on Field Food Crops in Africa and the Near East, Nairobi, 6–12 June.

————. 1988. *Response Farming in Rainfed Agriculture*. Davis, CA: WHARF Foundation Press.

Stewart, J. Ian, Robert M. Hagan, W. O. Pruitt, and Warren A. Hall. 1973. *Water Production Functions and Irrigation Programming for Greater Economy in Project and Irrigation System Design and for Increased Efficiency in Water Use.* Report 14-06-D-7329, Engineering and Research Center, Bureau of Reclamation, US Department of the Interior, Washington, DC.

Stewart, J. Ian, and Charles T. Hash. 1982. "Impact of weather analysis on agricultural production and planning decisions for the semiarid areas of Kenya." *Journal of Applied Meteorology* 21(4):477–494.

Sumra, Suleman. 1974. "A history of agriculture in Handeni District up to 1961." Seminar paper, Department of History, University of Dar es Salaam.

———. 1975a. "An analysis of environmental and social problems affecting agricultural development in Handeni District." MA thesis, University of Dar es Salaam.

———. 1975b. "Problems of agricultural production in *ujamaa* villages in Handeni District." Economic Research Bureau Paper no. 75.3, University of Dar es Salaam.

Sunseri, Thaddeus. 1997. "Famine and wild pigs: Gender struggles and the outbreak of the Majimaji war in Uzaramo (Tanzania)." *Journal of African History* 38:235–259.

Survey Division. 1956. "Tsetse and sleeping sickness." In *Atlas of Tanganyika,* 3rd ed., p. 12. Dar es Salaam: Department of Lands and Surveys.

Swantz, Marja-Liisa. 1985. *Women in Development: A Creative Role Denied?* London: C. Hurst and Co.

Swynnerton, R. J. M. 1955. *A Plan to Intensify the Development of African Agriculture in Kenya.* Nairobi: The Government Printer.

Tanzania Gender Networking Programme. 1994. *Structural Adjustment and Gender Empowerment or Disempowerment.* Dar es Salaam: Tanzania Gender Networking Programme.

Tanzania Ministry of Agriculture, Forestry, and Wildlife, Agricultural Division. 1965. *Report on the Development Possibilities of the Handeni Preserved Area.* Arnhem, Netherlands: published on behalf of the Tanzanian Ministry of Agriculture by International Land Management Consultants.

Taube, Guenther. 1989. "Stabilization and structural adjustment in Tanzania: The economic recovery programme, 1986–89, and its effects in the agricultural sector; the case of Lushoto District, Tanga Region." Lushoto. Mimeographed.

Taylor, D. R. F. 2004. "Capacity building and geographic information technologies in African development." In S. D. Brunn, S. L. Cutter, and J. W. Harrington, Jr., eds., *Geography and Technology.* Dordrecht, Netherlands: Kluwer Academic Publishers.

Thomas, Ian D. 1967. *A Survey of Administrative Boundary Changes in Tanganyika/Tanzania, 1957 to 1967, for Use in Intercensal Comparisons.* Dar es Salaam: Bureau of Resource Assessment and Land Use Planning, University of Dar es Salaam.

———. 1972. "Infant mortality in Tanzania." *East African Geographical Review* 10:5–26.

Thompson, J., I. T. Porras, J. K. Tumwine, M. R. Mujwahuzi, M. Katui-Katua, N. Johnstone, and L. Wood. 2004. *Drawers of Water II: Assessing Long-Term Change in Domestic Water Use in West Africa*. International Institute for Environment and Development, Sustainable Agriculture and Rural Livelihoods, SARL Project Summary. www.iied.org/sarl/research/ projects/t1proj05.html.

Thornthwaite, C. W. 1948. "An approach toward a rational classification of climate." *Geographical Review* 38:55–94.

Thornton, P. K., A. R. Saka, U. Singh, J. D. T. Kumwenda, J. E. Brink, and J. B. Dent. 1995. "Application of a maize crop simulation model in the central region of Malawi." *Experimental Agriculture* 31:213–226.

Tiffen, M. 1991a. *Environmental Change and Dryland Management in Machakos District, Kenya, 1930–90, Population Profile*. Working Paper 54. London: Overseas Development Institute.

———, ed. 1991b. *Environmental Change and Dryland Management in Machakos District, Kenya, 1930–90, Production Profile*. Working Paper 55. London: Overseas Development Institute.

———, ed. 1992. *Environmental Change and Dryland Management in Machakos District, Kenya, 1930–90, Institutional Profile*. Working Paper. London: Overseas Development Institute.

———. 1993. "Productivity and environmental conservation under rapid population growth: A case study of Machakos District." *Journal of International Development* 5:207–223.

Townsend, Meta K. 1998. *Political-Economy Issues in Tanzania: The Nyerere Years*. Lewiston, NY: Edwin Mellen Press.

Trapnell, C. G., and I. Langdale-Brown. 1969. "Natural vegetation." In W. T. W. Morgan, ed., *East Africa: Its People and Resources*. Nairobi: Oxford University Press.

Tripp, Aili Mari. 1994. "Gender, political participation and the transformation of associational life in Uganda and Tanzania." *African Studies Review* 37:107–131.

———. 1997. *Changing the Rules: The Politics of Liberalization and the Urban Informal Economy in Tanzania*. Berkeley and Los Angeles: University of California Press.

Turner, B. L., II. 1989. "The specialist-synthesis approach to the revival of geography: The case for cultural ecology." *Annals of the Association of American Geographers* 79:88–100.

———. 1997. "Spirals, bridges and tunnels: Engaging human-environment perspectives in geography." *Ecumene* 4:196–217.

———. 2002. "Contested identities: Human-environment geography and disciplinary implications in a restructuring academy." *Annals of the Association of American Geographers* 92:52–74.

Turner, B. L., II, and S. B. Brush, eds. 1987. *Comparative Farming Systems*. New York: Guilford Press.

Turner, B. L., II, Goran Hyden, and Robert W. Kates. 1993. *Population Growth and Agricultural Change in Africa*. Gainesville: University Press of Florida.

Turner, B. L., II, R. E. Kasperson, P. A. Matson, J. J. McCarthy, R. W. Corell, L. Christensen, N. Eckley, J. X. Kasperson, A. Luers, M. L. Martello, C. Polsky,

A. Pulsipher, and A. Schiller. 2003a. "A framework for vulnerability analysis in sustainability science." *Proceedings of the National Academy of Science* 100:8074–8079. http://www.pnas.org/cgi/content/full/100/14/8074.

Turner, B. L., II, P. A. Matson, J. J. McCarthy, R. W. Corell, L. Christensen, N. Eckley, G. K. Hovelsrud-Broda, R. E. Kasperson, J. X. Kasperson, A. Luers, M. L. Martello, S. Mathiesen, R. Naylor, C. Polsky, A. Pulsipher, A. Schiller, H. Selin, and N. Tyler. 2003b. "Illustrating the coupled human-environment system for vulnerability analysis: Three case studies." *Proceedings of the National Academy of Science* 100:8080–8085. http://www.pnas.org/cgi/content/full/100/. . . tspec=relevance&volume=100&firstpage=8080.

Turner, D. J. 1966. "An investigation into the causes of low yield in late-planted maize." *East African Agricultural and Forestry Journal* 32:249–260.

Uchara, G., et al. 1993. "The IBSNAT Project." In F. W. T. Penning de Vries et al., ed., *Systems Approaches for Agricultural Development.* Boston: Kluwer Academic Publishers.

United Nations Environmental Program. 2000. "Fanya-juu terracing." *Sourcebook of Alternative Technologies for Freshwater Augmentation in Africa.* UNEP. http://www.unep.or.jp/ietc/publications/techpublications/techpub-8a/fanya.asp.

United Republic of Tanzania. 1968. *1967 Population Census,* vol. 1. Dar es Salaam.

———. 1969a. *The Annual Economic Survey, 1968.* Dar es Salaam: The Government Printer.

———. 1969b. *Tanzania, Second Five-Year Plan for Economic and Social Development.* Vol. 2,*The Programmes.* Dar es Salaam: The Government Printer.

———. 1972. *The Economic Survey, 1971–72.* Dar es Salaam: The Government Printer.

———. 1976. *Tanga Water Master Plan, Tanga Region.* Vol. 5, *Socio-economics.* Essen: German Agency for Technical Cooperation.

———. 1992. *Joint Government of Tanzania/World Bank Agriculture Sector Review.* Dar es Salaam: Government Printer.

———. 2003. *Population and Housing Census, 2002.* http://www.tanzania.go.tz/censusdb/ageSexRegionAgeGroups.asp.

Vayda, A. P., and B. J. McCay. 1975. "New directions in ecology and ecological anthropology." *Annual Review of Anthropology* 4:293–306.

Vayda, A. P., and B. B. Walters. 1999. "Against political ecology." *Human Ecology* 27:167–179.

von Mitzlaff, U. 1988. *Women Farmers or Farmers' Wives?* Eschborn: Deutsche Gessellschaft für Techniche Zusammenarbeit.

Waddell, E. W. 1977. "The hazards of scientism: A review article." *Human Ecology* 5:69–76.

Walsh, S. J., T. P. Evans, and B. L. Turner II. 2004. "Population-environment interactions with an emphasis on land-use/land-cover dynamics and the role of technology." In S. D. Brunn, S. L. Cutter, and J. W. Harrington Jr., eds., *Geography and Technology.* Dordrecht, Netherlands: Kluwer Academic Publishers.

Wang'ati, F. J. 1968. "The water use of maize and beans—results obtained with hydraulic weighing lysimeters, Mwea irrigation scheme." Paper presented at Fourth Specialist Meeting on Applied Meteorology, Nairobi, 26–27 November.

————. 1969. "Methods of estimating photosynthesis in the field and their application in land use planning." *Proceedings, 4th Specialist Meeting on Applied Meteorology.* Nairobi: East African Agriculture and Forestry Research Organization.

————. 1972. "Lysimeter study of water use of maize and beans in East Africa." *East African Agricultural and Forestry Journal* 38:141–156.

Watts, M. 1983. *Silent Violence: Food, Famine, and Peasantry in Northern Nigeria.* Berkeley and Los Angeles: University of California Press.

White, G. F., ed. 1974. *Natural Hazards: Local, National, Global.* London: Oxford University Press.

White, G. F., D. S. Bradley, and A. U. White. 1972. *Drawers of Water: Domestic Water Use in East Africa.* Chicago: University of Chicago Press.

Wilbanks, T. J., and R. W. Kates. 1999. "Global change in local places: How scale matters." *Climate Change* 43:601–628.

Willatt, S. T. 1968. "Moisture use by tea in southern Malawi." Paper presented at Fourth Specialist Meeting on Applied Meteorology, Nairobi, 26–27 November.

Williams, C. C., and A. C. Millington. 2004. "The diverse and contested meanings of sustainable development." *Geographical Journal* 170:99–104.

Willis, Justin. 1992. "The making of a tribe: Bondei identities and histories." *Journal of African History* 33:191–208.

————. n.d. *A Guide to Places of Historical Interest in the Muheza Area.* Nairobi: British Institute in Eastern Africa.

Willis, Justin, and Suzanne Miers. 1997. "Becoming a child of the house: Incorporation, authority and resistance in Giryama society." *Journal of African History* 38:479–495.

Winans, Edgar V. 1962. *Shambala: The Constitution of a Traditional State.* London: Routledge and Kegan Paul.

Wisner, B. 1977. "The human ecology of drought in eastern Kenya." PhD diss., Clark University, Worcester, MA.

Woodhead, T. 1968. *Studies of Potential Evaporation in Tanzania.* Nairobi: East African Agriculture and Forestry Research Organization.

World Bank. 1992. *Agriculture Sector Review, Preliminary Findings.* Dar es Salaam: Government of United Republic of Tanzania and World Bank.

Wright, Marcia. 1968. "Local roots of policy in German East Africa." *Journal of African History* 9:621–630.

Young, R., and H. Fosbrooke. 1960. *Smoke in the Hills.* Evanston, IL: Northwestern University Press.

Yudelman, Montague, Annu Ratta, and David Nygaard. 1998. *Pest Management and Food Production.* Food, Agriculture, and the Environment Discussion Paper 256. Washington, DC: International Food Policy Research Institute.

Zimmerer, K. S. 1994. "Human geography and the new ecology: The prospect and promise of integration." *Annals of the Association of American Geographers* 84:108–125.

Zimmerer, K. S., and T. J. Bassett, eds. 2003. *Political Ecology: An Integrative Approach to Geography and Environment-Development Studies.* New York: Guilford Press.

Index

Page numbers in **boldface** refer to figures or tables.

University of Chicago Geography Research Papers

Titles in Print

222.   MARILYN APRIL DORN, *The Administrative Partitioning of Costa Rica: Politics and Planners in the 1970s,* 1989

217–18.   MICHAEL P. CONZEN, ed., *World Patterns of Modern Urban Change: Essays in Honor of Chauncy D. Harris,* 1986

216.   NANCY J. OBERMEYER, *Bureaucrats, Clients, and Geography: The Bailly Nuclear Power Plant Battle in Northern Indiana,* 1989

213.   RICHARD LOUIS EDMONDS, *Northern Frontiers of Qing China and Tokugawa Japan: A Comparative Study of Frontier Policy,* 1985

210.   JAMES L. WESCOAT JR., *Integrated Water Development: Water Use and Conservation Practice in Western Colorado,* 1984

209.   THOMAS F. SAARINEN, DAVID SEAMON, and JAMES L. SELL, eds., *Environmental Perception and Behavior: An Inventory and Prospect,* 1984

207–8.   PAUL WHEATLEY, *Nagara and Commandery: Origins of the Southeast Asian Urban Traditions,* 1983

206.   CHAUNCY D. HARRIS, *Bibliography of Geography,* part 2, *Regional,* volume 1, *The United States of America,* 1984

194.   CHAUNCY D. HARRIS, *Annotated World List of Selected Current Geographical Serials,* 4th ed., 1980

186.   KARL W. BUTZER, *Recent History of an Ethiopian Delta: The Omo River and the Level of Lake Rudolf,* 1971

152.   MARVIN W. MIKESELL, ed., *Geographers Abroad: Essays on the Problems and Prospects of Research in Foreign Areas,* 1973

132.   NORMAN T. MOLINE, *Mobility and the Small Town, 1900–1930,* 1971

127.   PETER G. GOHEEN, *Victorian Toronto, 1850 to 1900: Pattern and Process of Growth,* 1970